PLANT STRATEGIES AND THE

Dynamics and Structure of Plant Communities

MONOGRAPHS IN POPULATION BIOLOGY

EDITED BY ROBERT M. MAY

1. The Theory of Island Biogeography, by Robert H. MacArthur and Edward O. Wilson
2. Evolution in Changing Environments: Some Theoretical Explorations, by Richard Levins
3. Adaptive Geometry of Trees, by Henry S. Horn
4. Theoretical Aspects of Population Genetics, by Motoo Kimura and Tomoko Ohta
5. Populations in a Seasonal Environment, by Stephen D. Fretwell
6. Stability and Complexity in Model Ecosystems, by Robert M. May
7. Competition and the Structure of Bird Communities, by Martin L. Cody
8. Sex and Evolution, by George C. Williams
9. Group Selection in Predator-Prey Communities, by Michael E. Gilpin
10. Geographic Variation, Speciation, and Clines, by John A. Endler
11. Food Webs and Niche Space, by Joel E. Cohen
12. Caste and Ecology in the Social Insects, by George F. Oster and Edward O. Wilson
13. The Dynamics of Arthropod Predator-Prey Systems, by Michael P. Hassell
14. Some Adaptations of March-Nesting Blackbirds, by Gordon H. Orians
15. Evolutionary Biology of Parasites, by Peter W. Price
16. Cultural Transmission and Evolution: A Quantitative Approach, by L. L. Cavalli-Sforza and M. W. Feldman
17. Resource Competition and Community Structure, by David Tilman
18. The Theory of Sex Allocation, by Eric L. Charnov
19. Mate Choice in Plants: Tactics, Mechanisms, and Consequences, by Mary F. Willson and Nancy Burley
20. The Florida Scrub Jay: Demography of a Cooperative-Breeding Bird, by Glen E. Woolfenden and John W. Fitzpatrick

(list continues following the Subject Index)

PLANT STRATEGIES AND THE
Dynamics and Structure of Plant Communities

DAVID TILMAN

PRINCETON, NEW JERSEY
PRINCETON UNIVERSITY PRESS
1988

Published by Princeton University Press, 41 William Street,
Princeton, New Jersey 08540
In the United Kingdom: Princeton University Press, Oxford

Library of Congress Cataloging in Publication Data will be
found on the last printed page of this book

ISBN 0-691-08488-2 (cloth)
0-691-08489-0 (paperback)

This book has been composed in Linotron Baskerville

Clothbound editions of Princeton University Press books
are printed on acid-free paper, and binding materials are
chosen for strength and durability. Paperbacks, although satisfactory
for personal collections, are not usually suitable for library rebinding

Printed in the United States of America by Princeton University Press,
Princeton, New Jersey

9 8 7 6 5 4 3 2

FOR

Cathie
Lisa, Margie, and Sarah

Contents

Preface

The interactions between consumers and their resources, which can be a major determinant of patterns in nature, are strongly influenced by resource availabilities and by the foraging behavior of the consumers. Although it is common to think of the foraging behavior of animals, multicellular plants also have "foraging behaviors." A plant's ability to garner resources is strongly influenced by its morphology. Plant physiology and morphology interact to determine how growth depends on resource availabilities. A major advantage of plants, in addition to Harper's (1977) observation that they sit and wait to be counted, is that their above-ground morphology, and thus a major component of their foraging behavior, is visually obvious. Unfortunately, below-ground foraging effort is more difficult to observe. Plants have evolved a wondrous array of morphologies and life histories, and plant communities have many repeatable spatial and dynamic patterns. My desire to understand these was a major factor motivating this book. I started exploring these ideas more than two years ago with little idea where they would lead. I did start with the usual complement of prejudices and preconceptions, several of them highly cherished at the time, and found that some were reinforced and some rejected as I explored the logical implications of the mechanisms of competition for soil resources and light among size-structured plant populations.

Writing a book is a long, often tiring, and at times intellectually frightening journey, for there are many face-to-face encounters with the vast unknowns of our science. However, there are also exhilarating moments when disparate ideas coalesce, when patterns emerge from chaos. In looking back on the results of the past two years, I know that there is much more to be done. But the journey has produced insights into some of the fundamental processes, constraints, and tradeoffs that may have led to the broad, general patterns we see in the vegetation of the earth. I share these with you in this book. I do so in the spirit of one who knows that we have far to go before we truly understand nature. I hope that the ideas presented here may help guide you toward a better understanding of the forces shaping the evolution of plant traits and the structure and dynamics of plant communities.

This book could not have been written without the support, encouragement, and assistance of many. First and foremost, I thank my wife, Cathie, for her support during the all too frequent periods when writing led me to be distant and preoccupied. My next greatest debt is to Andrea Larsen, who prepared all the figures, corrected the text, prepared the bibliography, and assisted with almost all other aspects of manuscript preparation. Robert Buck and then Abderrahman El Haddi assisted with data analysis and with computer simulations. The ideas presented in this book have been influenced by many individuals with whom I have interacted over the years. I especially thank Nancy Huntly, Richard Inouye, John Pastor, Edward Cushing, John Tester, Eville Gorham, Hal Mooney, Margaret Davis, Peter Abrams, Lauri Oksanen, David Grigal, Deborah Goldberg, Norma Fowler, John Harper, Peter Vitousek, and David Wedin. I am deeply indebted to Hal Mooney, John Harper, Terry Chapin, John Pastor, Deborah Goldberg, Norma Fowler, Steve Pacala, Jim Grace, Scott Wilson, Nancy Johnson, Dave Wedin, Jim Grover, Jim Clark, Scott Gleeson, Steve Fifield, Barb Delaney, Bob McKane, Jenny Edgerton and a group of graduate students at The University of Michigan for their critical comments on the first draft of this book. I thank Judith May for her assistance in editing the manuscript. However, any and all errors that remain are mine. I thank the John S. Guggenheim Memorial Foundation for a fellowship that allowed me the time to start this work. I am greatly indebted to the Minnesota Supercomputer Institute for granting me time on a Cray 2. I also gratefully acknowledge support from the National Science Foundation (BSR-81143202 and BSR-8612104) for Long-Term Ecological Research at Cedar Creek Natural History Area, Minnesota. Without the support of NSF, the work presented here would not have been possible.

University of Minnesota, 1987

PLANT STRATEGIES AND THE Dynamics and Structure of Plant Communities

CHAPTER ONE

Introduction

> There is a very extensive literature in which it is demonstrated repeatedly that the [competitive] balance between a pair of species in mixture is changed by the addition of a particular nutrient, alteration of the pH, change in the level of the water table, application of water stress or of shading. These experiments had a significant historical importance in emphasizing that the interaction between a pair of species was a function of the environment in which the interaction occurred and an anecdotal value in defining, for a specialized condition of environment and species, the effects of a particular change. It is very doubtful whether such experiments have contributed significantly either to understanding the mechanism of "competition" or to generalizing about its effects.
>
> —J. L. Harper (1977, p. 369)

The central goal of ecology is to understand the causes of the patterns we observe in the natural world. The existence of patterns—of similarities from habitat to habitat—suggests that similar forces may have been at work in different habitats. This book is concerned with some of the broader, more general patterns that have been reported for terrestrial plants and with some of the forces that may have shaped plant morphologies, life histories, and physiologies, and thus determined the structure and dynamics of plant communities. Why is it, for instance, that species with similar physiological, morphological, and life history traits are dominant in a similar order during secondary succession in quite different habitats worldwide (Billings 1938; Bazzaz 1979; Christensen and Peet 1981; MacMahon 1981;

Cooper 1981; Inouye et al. 1987a; Tilman 1987a)? What causes primary successions in Indiana (Cowles 1899), Alaska (Crocker and Major 1955), and Australia (Walker et al. 1981) to be so similar, at least for their first 200 years? Within a geographical region, much of the variation in the species composition of vegetation is associated with the local soil type, especially the parent material on which the soil formed and the eventual productivity of the vegetation on that parent material in that climatic region (e.g., Lindsey 1961; Hole 1976; Rabinovitch-Vin 1979, 1983; Jenny 1980). Why is it that, both within and among biomes, species with similar maximal heights, relative growth rates, allocation patterns, and life histories tend to be dominant at similar points along such productivity gradients? Holding productivity constant, herbivory, disturbance, or other loss rates are a major determinant of vegetation composition (e.g., Lubchenco 1978; Grime 1979; Whitney 1986). Why, in a wide variety of habitats, does vegetation change along a gradient from low to high loss rates in qualitatively similar ways? An even more striking pattern that merits further explanation is the convergence of unrelated species to a common set of physiological, morphological and life history traits in widely separated but physically similar habitats worldwide (Mooney 1977; Cody and Mooney 1978; Orians and Paine 1983; Walter 1985). Further, almost all terrestrial vascular plants are alike in their modularity and their great morphological plasticity (Harper 1977).

The cause of such similarities is a central question facing plant ecologists. Might such similarities imply that terrestrial plant evolution and community structure have been greatly influenced by a few general underlying processes? Might a relatively simple approach be able to explain all these patterns on all these scales, at least in their broad outline? Or are such patterns unrelated to each other, with each pattern requiring a unique explanation?

INTRODUCTION

CONSTRAINTS AND TRADEOFFS

Pattern in ecology is caused by the constraints of the physical and biotic environment and by the tradeoffs that organisms face in dealing with these constraints. The more general and repeatable such constraints and tradeoffs are, the more general and repeatable will be the patterns caused by them. The long-term persistence of species requires that species be differentiated, i.e., that they have tradeoffs in their abilities to respond to the constraints of their environment. This book is concerned with the causes of broad-scale patterns of differentiation among terrestrial plants and the effects of such differentiation on the dynamics and structure of plant communities.

What, then, are the major constraints terrestrial plants face? Some general constraints on plants come from their place in food webs. All species are consumers of resources, some of which may be in short supply. Vascular plants require mineral nutrients, water, carbon dioxide, and light. Their abilities to use these resources depend on temperature, pH, humidity, and the oxygenation of the soil. In addition, plants are resources for a variety of species of herbivores, parasites, pathogens, and predators, and are also subject to loss and mortality caused by various disturbances to their habitat. Thus, within a given habitat, plants are constrained by resource availability and by loss or mortality caused by disturbance and herbivory.

Another constraint comes from the physical separation of essential plant resources. Terrestrial vascular plants require light, which is obtained above the soil surface, and mineral nutrients and water, which are obtained from the soil. Because these are nutritionally essential resources for photosynthesis, each plant requires a particular ratio of nutrient to light for it to have optimal growth. For a light-limited plant to obtain more light, it must allocate more of its growth to stems and leaves, and must allocate a smaller proportion of its growth to roots. Similarly, for a nutrient-limited plant to obtain relatively more nutrient, it must allocate more of its growth to roots, and thus proportionately less to leaves or stems. Thus, if a plant adjusts its allocation so as to increase its consumption of one of these resources, it necessarily decreases the relative amount of the other resource that it can acquire. This is an inescapable tradeoff for terrestrial plants that is dictated by their morphology and the physical separation of soil resources and light. One of the major predictions of the theory developed in this book is that this tradeoff has been a dominant cause of the patterns we see in natural plant communities. This occurs because each unique habitat—each unique pattern of soil resource and light levels—favors plants with a unique morphology, physiology, and life history. Thus, the physiognomy of the vegetation within a habitat should be strongly influenced by the forces that control the vertical light gradient and the levels of limiting soil resources. If some general, repeatable processes control patterns of resource availability, these would lead to general, repeatable patterns in plant evolution and community structure.

Productivity Gradients

Two major factors determine the availabilities of a limiting soil resource and light in a habitat: the rate of supply of the soil resource and the loss or mortality rate that plants experience. As discussed in Chapter 9, loss or mortality rates and soil resource supply rates could be correlated in natural habitats, and such correlation could be a further cause of natural patterns. However, it is instructive initially to consider the effects of each of these when the other is held constant. Holding loss or mortality rates constant, the habitat in which a plant lives can be classified as falling along a gradient from areas that have a low supply rate of a limiting soil resource, low soil resource levels, low plant biomass, and high penetration of light to the soil surface, to areas with a high supply rate of the soil resource, high soil resource levels, high plant biomass, and low penetration of light to the soil surface. For convenience, I will call such gradients "productivity gradients" or "soil-resource:light gradients." Light intensity at the soil surface is important because seedlings and shoots of newly establishing plants are short, and their growth rate is influenced by the light intensity they experience. This inverse correlation between the supply rate of a limiting soil resource and light intensity at the soil surface along productivity gradients is a major constraint of the terrestrial habitat.

Throughout this book, I will distinguish between resource levels and resource supply rates for soil resources. I define a resource level as the measurable concentration of the usable form or forms of a resource in the soil. I will, at times, use resource "availability" as synonymous with resource level. I define the supply rate of a resource as the rate at which usable forms of a resource are released into the soil. I do not define a supply rate for light because the canopies of all stands of vegetation receive full sunlight. Rather, I consider how the vegetation influences the vertical light gradient, especially light intensity at the soil surface.

Productivity gradients have been found to occur on a variety of spatial scales. For instance, the sandplains of Minnesota, Wisconsin, and Michigan, or the sandplains of northern Florida, have nutrient-poor soils, low standing crop, and high penetration of light to the soil surface, whereas soils formed on glacial till in Minnesota, Wisconsin, and Michigan have higher nutrient supply rates, higher plant biomass, and lower penetration of light to the soil surface. The differing parent materials of Blackhawk Island, Wisconsin, led to the development of soils that form a natural productivity gradient (Pastor et al. 1984). Further, all habitats have small-scale spatial variability in primary productivity and standing crop. Much of this variation may be caused by local differences in the soil resource supply rates.

Such local variation in soil resource supply rates can be caused by a variety of factors, including differences in soil permeability to water, exchange sites for nitrogen or phosphorus, effects of herbivore excretion, soil erosion, topographic variability, and feedback from plants (Jenny 1980). In some cases, productivity gradients occur as distinct gradients in space, such as the soil catenas that occur along slopes. However, soil-resource:light gradients need not be obvious gradients, spatially, but can exist wherever there is point-to-point variation in supply rates of limiting soil resources. Such gradients, on both large and small spatial scales, are likely to have been a major, general, repeatable feature of the habitat in which plants have evolved and differentiated.

Loss or Disturbance Gradients

A second major habitat constraint comes from disturbance, herbivory, predation, and other non-selective sources of loss of plant parts or mortality. For convenience, I will call all of these "loss" or "disturbance." Let's consider how the availability of a limiting soil resource and light availability at the soil surface would change along a hypothetical loss rate gradient—i.e., a gradient along which the loss rate changes but soil resource supply rates are held constant. Along such a gradient, there would be relatively low levels of both the soil resource and light at the soil surface in habitats with low loss rates. Habitats with high loss rates would have relatively high levels of the soil resource and of light at the soil surface. Thus, the levels of soil resources and of light at the soil surface should be positively correlated along a loss rate gradient but negatively correlated along a productivity gradient. Although there are fewer studies of loss or disturbance gradients than of productivity gradients, the examples show that soil resource and light levels are positively correlated on such gradients. Consider, for instance, the Hubbard Brook experiments (Likens et al. 1977; Bormann and Likens 1979). The undisturbed forest had high plant biomass, low penetration of light to the soil surface, and relatively low levels of extractable soil nutrients. Clear-cutting led to a great increase in light penetration to the soil surface and to a large increase in extractable soil nutrient levels, as indicated by the water leaching through the soil into the watershed. Nutrient levels increase following plant mortality because there is less plant biomass to consume nutrients as they become available (e.g., Vitousek et al. 1979, 1982). All forests and fields have natural disturbances which affect patches of various sizes within them. Averaging over disturbed and undisturbed patches within a whole forest or an entire field, the average level of extractable soil resources and of light at the soil surface should increase with the average loss rate (e.g., Swank et al. 1981; Vitousek et al. 1979, 1982; Vitousek and Matson 1985).

For simplicity, in this book I will combine all density-independent, non-selective processes causing death of plants or loss of plant biomass. I do this because, whatever the source of such loss or mortality, it should have a qualitatively similar effect on resource levels and thus a qualitatively similar effect on plant morphology and life history. Clearly, this is a major simplification which is only a first approximation for the effects of herbivores or various types of disturbances, both of which have selective, density-dependent components. I make this simplification to seek generality. However, there are many insights that would be gained from a more complex approach that included further details of the effects of specific herbivores or specific types of disturbances. These, though, are not the subject of this book. Throughout this book, I will use loss or disturbance interchangeably to refer to density-independent, non-selective losses that could be caused by herbivores, seed or seedling predators, fire, landslides, tree falls, and other processes.

The tendency for soil resource and light levels to be inversely correlated when loss rates are held constant is illustrated by a variety of field and greenhouse experiments. A visually clear example is provided by an elegant field study on the optimal nutrition of spruce trees in Sweden (Fig. 1.1; Tamm, Aronsson, and Burgtorf 1974; Tamm and Aronsson 1982; Tamm 1985). In reviewing nutrient addition experiments, Harper (1977) repeatedly emphasized that such experiments were difficult to interpret because "a major effect of supplying nutrients to vegetation may simply be to speed up the time at which light becomes limiting" (p. 340). In discussing a competition experiment between a grass and a clover (Stern and Donald 1962), Harper (1977, p. 361) said:

> The grass (adequately supplied with nitrogen) overtops the clover and the advantage is progressive, leading to the almost total suppression of the clovers. At first sight such an experimental result might have been interpreted as purely a problem in nitrogen nutrition. With no applied nitrogen the nodule-bearing and nitrogen-fixing legume was at an advantage—it evaded a struggle for existence for limiting nitrogen supplies. However, given adequate nitrogen the grass became the winner. Yet it is clearly unreal to separate the partitioning of nitrogen resources from the partitioning of incident radiation. The experiment starts as a single factor experiment but quickly turns itself into a study of the interactions between factors.

Harper is, indeed, correct that this interaction between soil resources and light complicates the design, implementation and interpretation of plant competition experiments. However, these are the underlying constraints and mechanisms of plant competition, and a mechanistic theory of plant competition should include them. The theory developed here in Chapters 3, 4, 6, and 7 models the vegetative growth of a plant as a continuous process that is determined by the pattern of allocation of photosynthate to additional stem, root, or leaf biomass. The model of nutrient and light competition among continuously growing size-structured plant populations that is developed in this book was designed to be as simple as possible and still include in it the major morphological and life history traits that influence the abilities of terrestrial plants to compete for soil

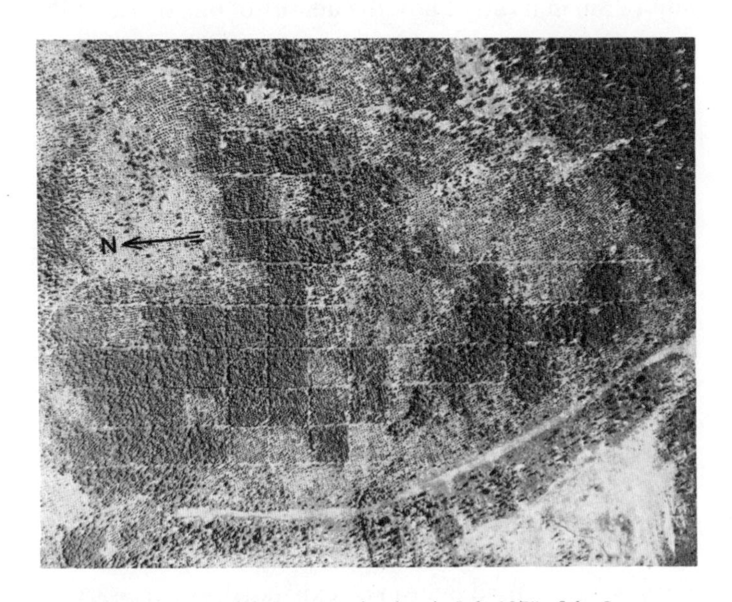

FIGURE 1.1. An aerial photograph taken in July 1975 of the Strasan, Sweden, experiments designed to determine the optimal nutrition of spruce trees. Nitrogen addition led to increased spruce biomass (the darker plots) and thus to decreased penetration of light to the soil surface. Nutrient addition began in 1967. Plots are generally 30 m × 30 m. Optimal growth occurred with the addition of N, P, K and Mg and was related to the nutrient content of the needles and the pattern of allocation to roots, needles, and stems (Tamm 1985). See Tamm et al. (1974) and Tamm (1985) for further details. The aerial photograph is reprinted from Tamm (1985) in the *Journal of the Royal Swedish Academy of Agriculture and Forestry*, Supplement 17, page 12, with the permission of the journal and the author. I thank Professor Carl Tamm for providing me with the original photograph and allowing me to reproduce it here.

resources and light. This model is used to explore a variety of questions about the evolution of plant morphologies (allocation patterns) and life histories, and about the effects of these plant traits on the dynamics and equilibrium structure of plant communities. The central goal of this book is to explore the logical implications of the mechanisms of plant competition for nutrients and light.

Most ecological theory has been phenomenological. It has described interactions such as competition or mutualism in terms of how the density of one species influences the growth rate of another species, without ever stating the actual mechanisms whereby one species influences the other. Such theory cannot explore the ramifications of these mechanisms for the evolution of species traits or for the structure and dynamics of populations, communities, and ecosystems (Tilman 1987a). This book takes a different approach—an approach that explicitly states the processes whereby individuals of one species influence individuals of that and other species. It is these mechanisms that have shaped the morphology, physiology, and life histories of species, and that have influenced the types of conditions for which each is dominant or rare. If there is to be any simplicity or generality in ecology, it will be found in environmental constraints and in the mechanisms of interaction, not in simple theories that ignore mechanisms. A major advantage of a mechanistic approach is that it can initially be narrowly focused but can be expanded, as necessary, to include other mechanisms and a larger portion of the foodweb and abiotic environment.

This book is limited in scope. It focuses on a few fundamental mechanisms of intraspecific and interspecific resource competition among terrestrial plants and the implications of these mechanisms for the evolution of plant traits and the dynamics and structure of plant communities. It does not treat plant-herbivore interactions, except the component mentioned above. This is done not to downplay the possible importance of herbivory, but to explore the logical consequences of the mechanisms of resource competition. Nor does this book explicitly consider the effects of neighbor-to-neighbor spacing in plant competition, a question addressed by Pacala and Silander (1985) and Pacala (1986). The underlying mechanisms of soil nutrient supply and the feedback effects of plants on soils are also not treated in depth. Each of these is an important area that merits further exploration and integration with the ideas presented here. However, no single study or book can encompass the full breadth of ecology. The natural, ecological world is phenomenally complex. Every insight—every hint of major underlying processes that structure it—is a hard-won advance. We now know that it is not a matter of competition versus predation versus mutualism versus disturbance as being "important" processes structuring the natural world (Quinn and Dunham 1983). All of these are important, and all must interact. However, there is still much to be gained by taking a simple perspective, and exploring the implications of a few factors, with other potentially important factors "held constant" for the sake of ease of analysis. Each advance thus gained provides the opportunity for a synthesis with advances that have been gained by making other simplifying assumptions.

The actual mechanisms of intraspecific and interspecific competition among multicellular plants are not simple. Multicellular plants have size and age dependent processes that greatly complicate any attempt to understand them. Plants are morphologically and physiologically plastic. They have a modular morphology that is composed of fairly fixed subunits (leaves, seeds, roots, stems; Harper 1977), but plants are capable of modifying, both phenotypically and genotypically, the relative allocations to these subunits. As will be discussed in Chapters 2 and 9, such morphological plasticity can influence the intraspecific and interspecific competitive ability of plants. Given such complexity, how should plant competition be viewed? One approach would be to ignore mechanisms and seek simplicity by describing the phenomenon of competition. This is just what is done by the Lotka-Volterra equations. Similarly, de Wit (1960) suggested that plant competition was so complex and so unique to each organism and habitat that there was no hope of formulating a general, mechanistic theory of plant competition. Instead, he suggested a phenomenological approach. In contrast, the approach taken in this book is to develop a simplified, but mechanistic, theory of competition that can be used to explore some of the general features of plant competition.

Mechanistic approaches impose a discipline and a limitation to vision that may be of great help in plant ecology. Some of the classical debates in plant ecology—such as the debate over whether or not a plant community exists as an entity in its own right or is an assemblage of species with individualistic responses—have occupied the efforts of far too many ecologists for far too many years. Such questions seem of trivial importance when a mechanistic approach is taken. A mechanistic approach eliminates the need to test among what may be spurious, broad-scale community generalizations that are not based on the evolutionary forces that have shaped plants, and instead focuses attention on more quantitative patterns that are predicted from underlying tradeoffs in the biology of the species and the constraints of the physical and biotic environment.

RESOURCE LIMITATION

A critical step in trying to understand plant competition is to determine what the actual mechanisms of interspecific interaction are for any given situation. The two most likely mechanisms of plant competition are resource (or exploitative) competition and allelopathic competition. Exploitative competition occurs when one plant inhibits another plant through consumption of limiting resources. Allelopathic competition occurs when one individual releases a compound that in some way inhibits growth or increases mortality of other plants. Neither of these mechanisms represents a *direct* effect of the density of one species on the growth rate of another species. In both cases, the density of each species directly influences some intermediate entity, and it is that entity that actually affects the growth rate of the other species. Thus, in order to demonstrate that species actually compete for resources in nature, it is necessary to manipulate experimentally resource levels in the field. Similarly, to demonstrate allelopathic competition in the field, it is necessary to manipulate the levels of allelopathic compounds in the field. The theory developed in this book applies only to cases of resource competition. Before any of this theory can be applied to a particular community, field experiments must be performed demonstrating that the plants are competing for resources and demonstrating which resources are limiting. Until this is done, it would be easy to gather data that seemed to support or refute this theory independent of the potential validity of the theory to that field situation.

The strength of a mechanistic approach to plant communities is that it can make explicit predictions about a wide range of patterns and processes in nature. However, a mechanistic theory can be misapplied just as easily as any other theory. A consistency between the predictions of a mechanistic theory of competition for nitrogen and light, and patterns seen in nature, for instance, is of no importance if nitrogen and light are not limiting in that community. To invoke such consistency without evidence of limitation is a potentially great danger. In this book, I will build a case on existing evidence as to the plausibility of such mechanisms as explanations for patterns we see in nature, but I wish to stress that most such cases merely demonstrate plausibility. I present them to encourage others to test the ideas developed here, not as a statement of "proof" of the underlying theory.

Finally, I should note that the approach taken in this book is conceptually quite different than that taken in Grime (1979) and leads to many conclusions that often directly contradict Grime's. Although I disagree with the ways in which he suggests various processes interact, I share with Grime (1979) the view that competition, loss rates (Grime's "disturbance"), and resource availabilities (with low availabilities being a major component of Grime's "stress") have a major influence on plant community structure. I will discuss the similarities and differences between Grime's perspective and mine as relevant throughout this book.

A PREVIEW

This book starts by using the equilibrium, resource-dependent growth isocline approach to competition developed in Tilman (1980, 1982) to demonstrate that the long-term average availability of a limiting soil nutrient and of light at the soil surface should depend on the nutrient supply rate and the loss rate of a habitat (Chapter 2). Because plants require both an above-ground resource (light) and below-ground resources (nutrients and water), plants face a tradeoff. To acquire more of one resource necessarily means that they must acquire proportionately less of another. Thus, the pattern of plant allocation to above- versus below-ground structures should influence the competitive ability of a plant in a given habitat (Chapters 4 and 5). However, all allocation to such non-photosynthetic structures as stems and roots necessarily decreases the maximal rate of vegetative growth of a plant (Chapter 3) and can thus greatly influence plant population dynamics (Chapter 6). The transient population dynamics that occur because of differences in maximal growth rates may be a major cause of the pattern of secondary succession, and may make it difficult to interpret the results of many short-term field experiments (Chapter 7). A five-year experimental study of plant distributions and successional dynamics at Cedar Creek Natural History Area, Minnesota, provides a wealth of information, much of it previously unpublished, with which to evaluate the predictions of the theory developed in this book (Chapter 8). The book ends with an exploration of some additional implications of the theory and with suggestions for further research (Chapter 9).

CHAPTER TWO

The Isocline Approach to Resource Competition

The complexities caused by the size structure of plant populations, by the linkage of nutrients and light, and by the tradeoff plants face in foraging for a limiting soil resource and light, mean that no simple theory can include all the components of plant competition. Does this necessarily mean that simple approaches are of no use? Complex models can often have much of their dynamic complexity adequately summarizable by a few equations (Schaffer 1981). In many complex processes, a few steps become rate limiting and thus become the major determinants of the patterns observed. The study of the mechanisms of plant competition is too young for us to know all the advantages and disadvantages of simple versus complex models. In this chapter I summarize a simple theory of plant competition for resources. In the remainder of this book, I develop a more complex and thus more realistic model of the mechanisms of plant competition, and compare its predictions with those based on the simpler theory developed in this chapter.

The simple theory uses the conditions that exist once each population reaches equilibrium to predict the outcome of interspecific competition for resources. This assumes that the resource requirements of various stages in the life cycle of each species can be summarized by their effect on the equilibrial resource requirements of that species. For populations that do not reach an equilibrium, I assume that long-term average resource availabilities may be a suitable approximation to equilibrial conditions. The equilibrial requirements are given by the resource-dependent growth isoclines of the species (Tilman 1980, 1982). Although much of the material in this chapter repeats earlier discussions in Tilman (1980, 1982), I also use this section to develop four basic concepts: (1) that the availabilities of all limiting resources should be positively correlated along loss rate gradients; (2) that the availability of a limiting soil resource and light should be negatively correlated along productivity gradients; (3) that plants should become separated along such gradients in a manner determined by their resource requirements; and (4) that optimal foraging of morphologically plastic plants for nutritionally essential resources should lead to a curved resource-dependent growth isocline, with morphologically plastic plants often being superior competitors compared to plants with a fixed morphology. Although I try to minimize the repetition of material that was published in Tilman (1980, 1982), some repetition is necessary. Those familiar with the earlier work may find it best to skim this chapter. I have tried to write it so that those with no familiarity with Tilman (1980) or Tilman (1982) may also understand it. The simple models developed in this chapter may be contrasted with a more complex model of allocation and growth for size-structured plants presented in Chapters 3 and 4. That model assumes that plant growth is determined by the pattern of allocation to roots, leaves, stems, and seeds. Each plant species is described by its allocation pattern, its seed size, its height at maturity, its maximal photosynthetic rate, the nutrient and light dependence of photosynthesis, the respiration rates of its different tissues, allometric and structural constraints, and other parameters.

COMPETITION FOR A SINGLE LIMITING RESOURCE

Let us first consider a case in which several different species are all limited by the same resource. What should be the outcome of interspecific competition, assuming that the interactions eventually lead to an equilibrium? In order to predict the outcome of competition for a single limiting resource, it is necessary to know the resource level at which the net rate of population change for a species is zero. This occurs when vegetative growth and reproduction balance the loss rate the species experiences in a given habitat. I call the resource level at which this occurs the R^* of that species for that limiting resource in that habitat. There are two distinct ways in which this information could be obtained. The first, and probably better, way to determine the R^* of a species in a given habitat would be to allow the species to attain its equilibrial biomass in a monospecific stand in that habitat. The level to which the species reduced the limiting resource at equilibrium would be its R^*. At equilibrium, the rate of resource consumption would equal the rate of resource supply. If a species were in a habitat in which the actual resource level was greater than R^*, the population size (by which I mean its mass per unit area once a stable age or size distribution was attained) would increase, thus reducing the resource level down toward R^*. If the resource level were less than R^*, population size would decrease, allowing the resource level to increase because of decreased consumption rates. It is only for habitats in which resource levels are at R^* that population size should remain constant. I call R^* the requirement of a species for a limiting resource.

The second way to determine R^* would be to determine the dependence of the growth rate of the species on resource levels, as illustrated by the resource-dependent growth curves of Figure 2.1. The y-axis of these figures is the long-term specific rate of growth or loss for the population ($dB/dt \cdot 1/B$, where B is biomass per unit area). If the population were a size-structured population (with population size expressed as biomass per unit area), the growth rate would be the natural logarithm of the dominant eigenvalue of the population projection matrix determined at each resource level, but in the absence of resource-independent mortality or other losses (e.g., Hubbell and Werner 1979; Vandermeer 1980). The total loss rate experienced by the population must be calculated in a comparable manner. For a size-structured population, this loss rate would be measured as the natural logarithm of the dominant eigenvalue of the population projection matrix that included all resource-independent loss terms, but no resource-dependent growth terms. The environmental availability of the resource at which the gain (from reproduction and vegetative growth) just balanced loss (from disturbance, herbivores, predation, and other mortality sources) would give R^* (Fig. 2.1).

When several species are all limited by a single resource, the one species with the lowest R^* is predicted to competitively displace all other species at equilibrium (O'Brien 1974; Tilman 1976; Hsu et al. 1977; Armstrong and McGehee 1980). The mechanism of competitive displacement is resource consumption. The population size of the species with the lowest R^* should be able to continue increasing until that species reduces the resource level (concentration) down to its R^*, at which point there would be insufficient resource for the survival of the other species. Several experimental tests of the R^* criterion of competition for a single limiting resource are summarized in Tilman (1982). One possible theoretical case is illustrated in Figure 2.1.

The theory presented here suggests that plants should compete strongly in habitats with low resource levels. This view contrasts with the assertion made by Grime (1979) that plants do not compete when they live in either "stressed" environments, such as low-nutrient environments, or in habitats with high disturbance rates. Grime's assertion, though, is inconsistent with numerous studies of intraspecific and interspecific competition. If plants did not compete on nutrient-poor soils, then, when growing in mono-

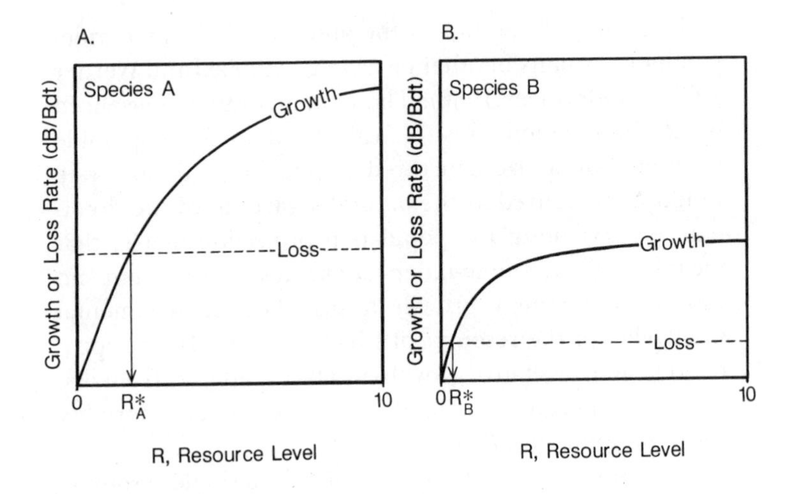

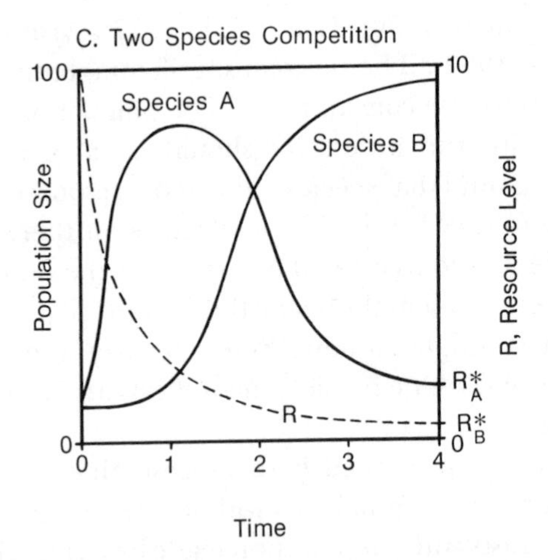

FIGURE 2.1. (A) Resource-dependent growth (solid curve) and loss (broken line) for species *A*. R_A* is the amount of the resource species *A* requires to survive in this habitat. (B) Similar curves for species *B*. (C) When two species compete for a single limiting resource (*R*), species *B*, which has the lower equilibrial resource requirement (*R**), should completely displace species *A* once equilibrium is reached.

culture, plant relative growth rates (*dB/Bdt*, where *B* is plant biomass) and average weight per plant would not decrease with increases in plant density on such soils. However, many studies have shown that relative growth rates and weight per plant decrease with increases in initial plant density on both poor and rich soils (e.g., Donald 1951; Clatworthy 1960; Harper 1961, 1977). Indeed, this decrease in the growth rate of individual plants with increases in plant density is a prerequisite for the "law" of constant final yield (Kira et al. 1953; Harper 1977). Cowan (1986) grew 8 herbs at both high and low densities along a gradient ranging from extremely nitrogen-poor subsurface sands to rich prairie soils. Along this full gradient, the final average weight per plant (Fig. 2.2) and the relative growth rate were significantly lower at high plant densities for each of the species, demonstrating strong intraspecific competition on all these soils, including the extremely nitrogen-poor soils. Inouye et al. (1980) and Inouye (1980) showed strong competition among desert annuals. Stern and Donald (1962) showed that clover displaced a grass from a nitrogen-poor soil after 133 days of competition. This competitive outcome was reversed in plots receiving high rates of nitrogen addition. Mahmoud and Grime (1976) studied competitive interactions among all pairs of three grass species (*Arrhenatherum elatius*, *Festuca ovina*, and *Agrostis tenuis*) on a rich soil and a poor soil. In comparison to the monocultures, the presence of a second species led to decreased weight per plant on their poor soil, with *Agrostis* causing a 30% decrease in the weight/shoot of *Arrhenatherum* and a 43% decrease for *Festuca*, and *Arrhenatherum* causing a 59% decrease for *Agrostis* in the low nitrogen treatment. Thus, there is experimental evidence showing both intraspecific and interspecific competition on nutrient-poor soils, as predicted by the theory summarized above.

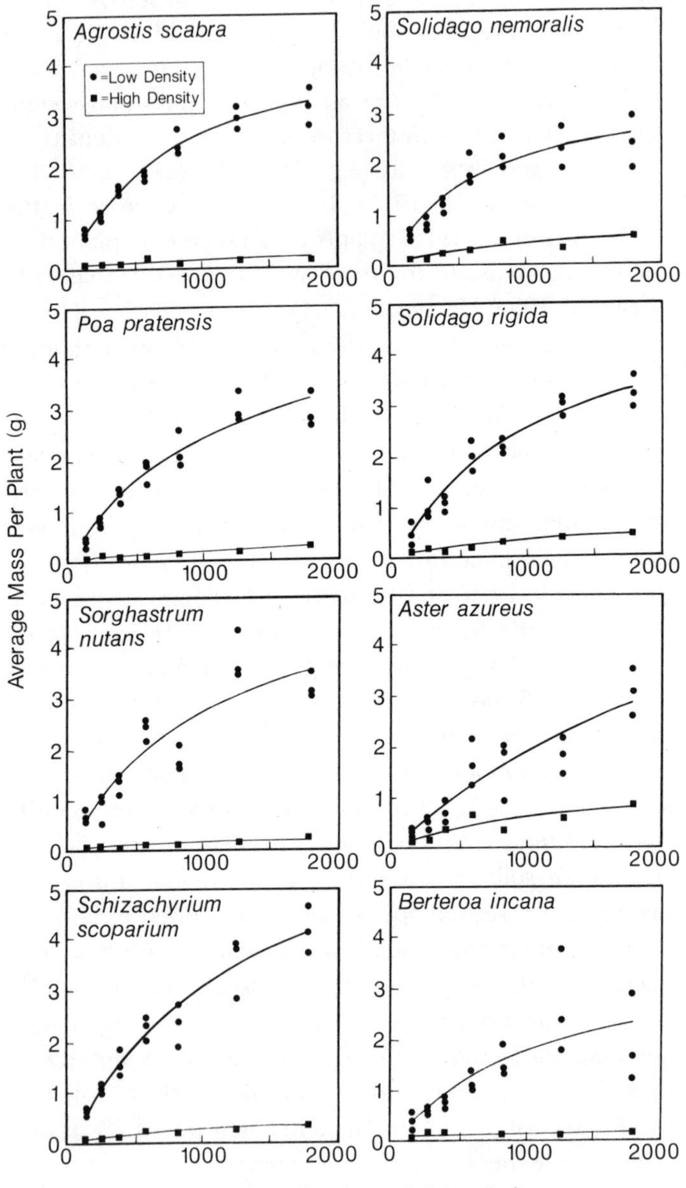
Agrostis scabra
●=Low Density
■=High Density
Solidago nemoralis
Poa pratensis
Solidago rigida
Sorghastrum nutans
Aster azureus
Schizachyrium scoparium
Berteroa incana
Average Mass Per Plant (g)
Total Nitrogen in Soil (mg/kg)
0
1
2
3
4
5
0
1000
2000

R^* AND LOSS OR DISTURBANCE RATES

The eventual level to which a limiting resource is reduced at equilibrium is determined by the resource dependence of growth and by loss. The loss rate of a population is caused by numerous components, including disturbance, seed predation, and herbivory. Independent of the causes of such losses, of the number of species competing for a resource, or the competitive abilities of the species in a particular habitat, average (equilibrial) resource levels (R^*) will increase with the loss rate (Fig. 2.3). This increase in R^* with the loss rate is a direct result of the dependence of growth rate on resource level. For equilibrium to occur, growth must balance loss. At equilibrium, higher loss rates must be accompanied by higher growth rates, and higher growth rates require higher resource levels. Thus, average resource levels must increase with increases in the loss rate (Fig. 2.3A). These higher resource levels are accompanied by lower consumer biomass at the higher loss rates.

This pattern holds no matter how many species are competing, and no matter how their growth rates depend on resource levels, if growth rates increase with resource availability and if loss rates of all species increase. Consider a case in which species with higher maximal growth rates have lower growth rates in resource-poor habitats. As loss rate increases, equilibrial resource levels will increase and

---

FIGURE 2.2. Average mass per plant for 8 species of old field plants grown outdoors in 18-liter pots at either high density (about 100 plants per pot) or at low density (7 plants per pot) along a nitrogen gradient. The nitrogen gradient was established by mixing a rich surface soil with different proportions of a subsurface sand. Nitrogen mineralization rates were highly correlated with total soil nitrogen. At all levels of nitrogen, average mass per plant was lower in the high density pots than the low density pots, demonstrating that intraspecific competition occurs throughout the full range of nitrogen availabilities. The poorest soils (total N of 125 mg/kg) are poorer than any naturally occurring soils ever sampled in the old fields of Cedar Creek. The richest soils (1825 mg/kg) are similar to the richest soils found at Cedar Creek. From Cowan (1986).

there will be a sequence of species replacements (Fig. 2.3B). If species do not have such tradeoffs (Fig. 2.3C), a single species will be dominant at all loss rates, but R^* will still increase with the loss rate. Equilibrium resource levels will increase with loss rates, whatever their source, as long as all species experience the increased loss, even if they do not all experience it equally. Although the shape of the curve between equilibrium resource availability and loss rate is determined by many factors, equilibrial resource levels must be an increasing function of the loss rate. The relationships of Figure 2.3C lead to curve 1 of Figure 2.3D and those of Figure 2.3B lead to curve 2 of Figure 2.3D.

This tendency for equilibrial resource levels to increase with the loss rate of a habitat is a major element providing structure to habitats. Further, it illustrates the inseparable link between competition and all sources of loss, including disturbance and herbivory. Competition does not become unimportant, as many have argued (e.g., Grime 1979; Harris 1986; Wiens 1984), in habitats with high loss rates. While it is true that such habitats may have high levels of all limiting resources, species require high resource levels to survive in these habitats. If plants that are competing for resources are also subject to herbivory, it is logically incorrect to say that herbivory is important and competition is unimportant in habitats with high herbivore densities (Quinn and Dunham 1983). All processes that influence the loss rates of species necessarily also influence their competitive interactions. Competition cannot be separated from disturbance or herbivory. It is this link between loss and growth that establishes one of the major gradients along which plants may have differentiated. This is a gradient from habitats with low loss rates and low equilibrial levels of all resources to habitats with high loss rates and higher equilibrial levels of all resources. This same qualitative pattern holds true even for oscillatory or fluctuating habitats that never reach equilibrium: the long-term average levels of all

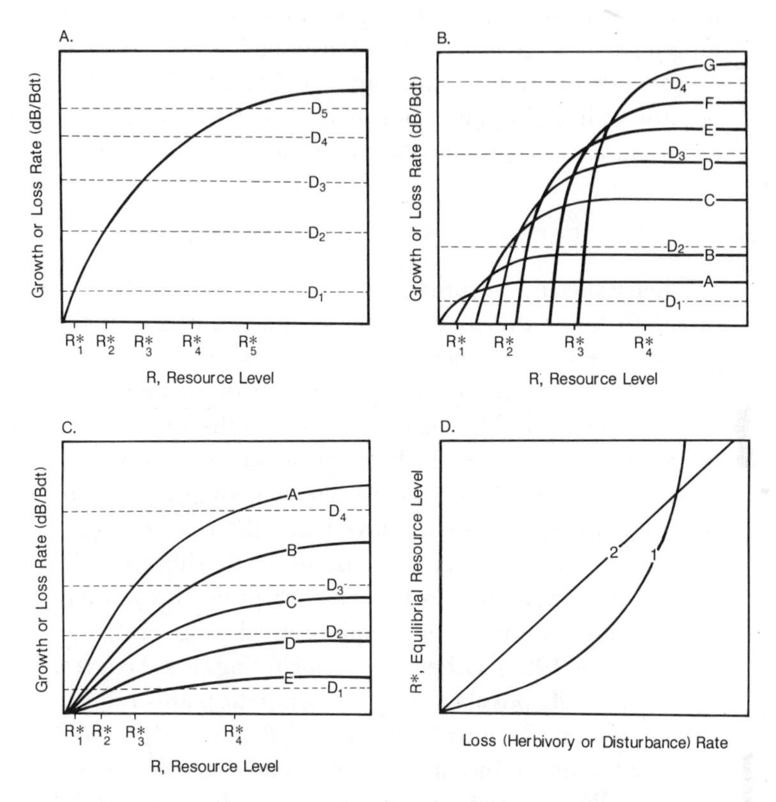

FIGURE 2.3. (A) Effect of loss rate on average resource availability. The solid curve gives the dependence of dB/Bdt on resource availability. D_1 through D_5 (shown with dotted lines) are different disturbance, death, herbivory, or other loss rates. The R^*'s show the equilibrial resource availabilities associated with each loss rate. A population can be maintained in a habitat only if its growth rate can at least balance its loss rate. Because growth rate (dB/Bdt, where B is biomass) increases with resource availability (R), for a population to survive at higher loss rates it must have higher resource availabilities. (B) R^* increases with loss rates even when many different species are competing. Species A should displace all other species for D_1, species C should win for D_2, species E should win for D_3, and species G should win for D_4. (C) R^* increases with D in a case in which species A is always a superior competitor. (D) These cases show that R^* must always increase with the loss or disturbance rate. For the case in part B, R^* would increase almost linearly with D (curve #2). For the case of part C, R^* would increase almost exponentially with D (curve #1).

limiting resources should be higher in habitats with higher long-term average loss rates. By "average" I mean both an average through time and an average through space, because individual plants and their sources of mortality or loss occur at discrete points in space and time.

RESOURCE-DEPENDENT GROWTH ISOCLINES

When a species consumes two or more resources, it is necessary to know the total effects of the resources on the growth rate of the species. These effects can be summarized, at equilibrium, by the zero net growth isocline of the species (Tilman 1980). This isocline shows the levels of two (or more) resources at which the growth rate per unit biomass of a species balances its loss rate. The shape of the isocline of a species can be used to define the resource type. Thus, a pair of resources may be perfectly substitutable, complementary, antagonistic, switching, perfectly essential, interactively essential, or hemi-essential (Fig. 2.4; see Tilman 1980, 1982). Other shapes, including closed curves, are theoretically possible. All photosynthetic plants require light, water, and various forms of C, N, P, K, Ca, Mg, S, and about 20 other mineral elements. These resources are nutritionally essential with respect to each other. Many animal resources tend to be substitutable or switching, which leads to some interesting differences between plant and animal competition, and potentially, their diversity patterns (Tilman 1982). If two resources are perfectly essential, an increase in one resource cannot overcome limitation

FIGURE 2.4. The solid curve in each figure is the zero net growth isocline of a population. A zero net growth isocline shows the availabilities of resources 1 and 2 for which growth and reproduction exactly balance all sources of loss. Note that R_1 and R_2 are resource levels, i.e., the measurable concentration of the usable form of each resource in a habitat. Population size would increase for resource levels in the shaded regions and decrease for levels in the unshaded region.

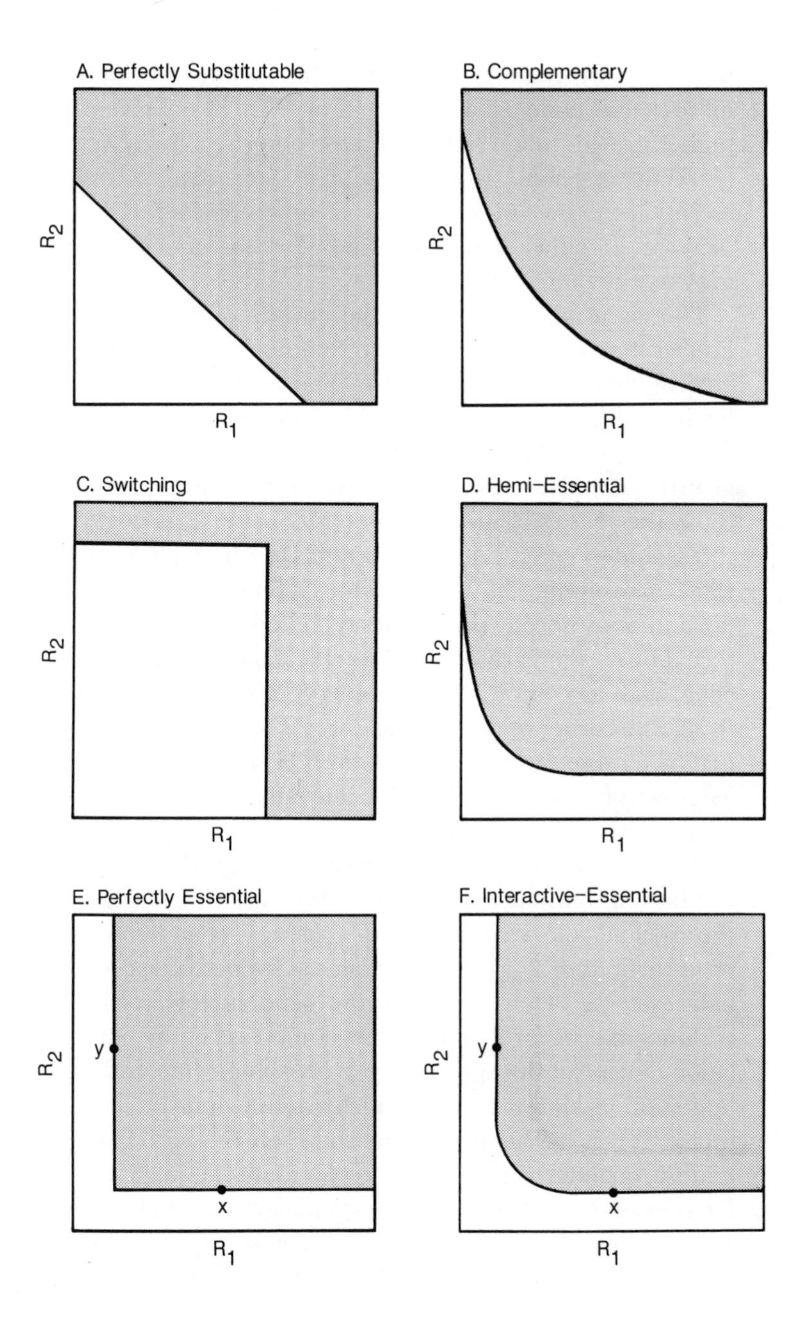

A. Perfectly Substitutable
R2
R1
B. Complementary
R2
R1
C. Switching
R2
R1
D. Hemi-Essential
R2
R1
E. Perfectly Essential
y
x
R2
R1
F. Interactive-Essential
y
x
R2
R1

by another (Fig. 2.5A). Thus, if a plant is limited by light, increased levels of available P or K or N could not give it a higher growth rate if light and these other resources were perfectly essential. If two resources were interactively essential, the isocline would have a curved corner and an increase in either resource could lead to an increased growth rate (Fig. 2.5B).

The right angle corner for the isocline of Figure 2.4E implies that R_1 and R_2 are perfectly essential. If a habitat has environmental availabilities of R_1 and R_2 that fall on the horizontal portion of the isocline, such as at point *x*, increasing or decreasing R_1 would have no effect on the growth rate of this species. However, increasing R_2 would cause the species to increase in density because its growth rate would be greater than its loss rate. Decreasing R_2 would cause it to decline in density. Thus, along the horizontal portion of its isocline this species is limited by R_2 and unaffected by R_1. Similarly, along the vertical portion of the isocline, such as point *y*, it is limited by R_1 and unaffected by R_2. At the corner of the isocline (Fig. 2.5A), it is jointly limited by both resources, and both must be increased to have its growth rate increase, but only one must be decreased to have its growth rate decrease.

If two resources are interactively essential, the isocline will have a rounded corner (Fig. 2.4F). At points *x* or *y* on the isocline, the response of this species to addition or removal of R_1 or R_2 is almost the same as for perfectly essential resources. However, near the bend in the isocline, adding either R_1 or R_2 can lead to an increase in the population density of this species (Fig. 2.5B). Thus there is dual limitation by both resources, which are functionally substitutable for each other over a range of resource levels. There is little qualitative difference, ecologically, between perfectly essential and interactively essential resources (Tilman 1982). None of the ideas presented here or previously would be any different, qualitatively, if they were based on

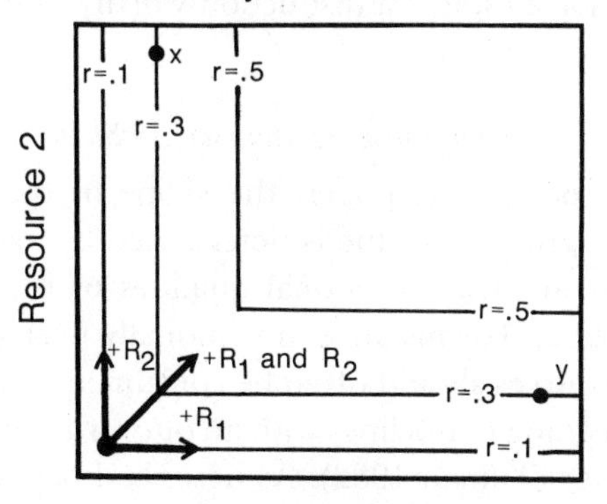

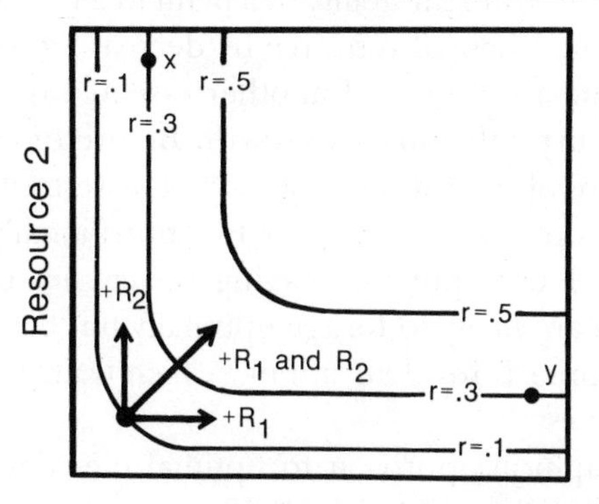

FIGURE 2.5. (A) Resource-dependent growth isoclines for a population with perfectly essential resources. (B) Isoclines for a population with interactively essential resources. The three isoclines shown in each part give the resource levels at which the per capita reproductive rate, r, would be 0.1, 0.3, or 0.5 time^{-1}. The arrows show the effect on reproductive rate of adding just R_1, just R_2 or adding both ($R_1 + R_2$) resources.

interactively essential versus perfectly essential resources. However, because optimal foraging theory predicts, as discussed below, that multicellular plants should have rounded isoclines, it is a distinction worthy of experimentation.

Optimal Foraging and Isocline Shape

For any pair of resources, the shape of the resource-dependent growth isocline is determined by many factors other than just the nutritional qualities of the resources (Tilman 1982). For instance, nutritionally perfectly substitutable resources should often be consumed in a switching manner, leading to isoclines with an outward-bowing right-angle corner (Tilman 1982). As described below, a simple theory of optimal foraging for nutritionally essential resources predicts that, in three of four cases, the resulting resource-dependent growth isoclines should have a rounded corner, not a right-angle corner. The rounded corner comes from the ability of a plant to increase its acquisition of one essential resource by decreasing its expenditure for the acquisition of another essential resource that does not currently limit its growth. Before discussing the implications of this for the evolution of terrestrial plants, let us discuss optimal foraging for two nutritionally essential resources. I use optimal foraging theory not because all organisms are likely to forage optimally but because it is a simple approach for determining which traits may lead to highest fitness.

The graphical approach to optimal diet developed by Rapport (1971) and Covich (1972) allows many aspects of optimal diet to be easily explored. This approach considers two constraints on diet: constraints on the amounts of two resources that an individual can consume per unit time (called the consumption constraint) and constraints on the fitness of the individual that depend on the nutritional aspects of the resources and amounts of the two resources consumed (called the nutritional constraint). For nutritionally perfectly essential resources, the nutritional constraint can be represented by isoclines of equal fitness (growth rates), such as the curves numbered 1 and 2 in Figure 2.6. The right-angle corner is caused by the resources being nutritionally perfectly essential. The nutrition isoclines that are further from the origin are curves of higher fitness, indicating that growth rate increases with the amount of the resources consumed per unit time. (Note that the axes for these figures are the amounts of R_1 and R_2 consumed, not their resource levels.) The consumption constraint curve (labeled *a* or *b* in Fig. 2.6) represents the maximal amounts of two resources that an individual can consume in a given amount of time. There are four possible shapes for consumption constraint curves. Three of these shapes (linear, convex, and concave) are based on the assumption that there is a tradeoff in foraging for one resource versus the other. For these three consumption constraint curves, an increased consumption of one resource can only be accomplished by decreased consumption of the other. For plants, such a tradeoff would be reflected in morphological or physiological plasticity. For instance, plants that have a higher proportion of their biomass in leaves and stems necessarily have a lower proportion in roots. The actual shape (linear, convex, or concave) of the consumption constraint curve is determined by physiological, morphological and environmental factors. The important aspect of all three of these is that they assume plasticity in plant foraging, i.e., that a plant can acquire more of one resource by acquiring less of the other. In contrast, the consumption constraint curves with a perfect right-angle corner (Fig. 2.6C and E) assume that the plants are not plastic, but have independent, fixed abilities to consume the two resources.

The optimal diet of an individual is that point on its consumption constraint curve that leads to the highest fitness. This is the point of contact between the consumption constraint curve and the nutritional constraint curve that has the highest fitness (is furthest from the origin) but still touches the consumption constraint curve (i.e., is a possible diet). For linear, convex, or concave consumption constraint curves, the optimal diet is always consumption of the two resources in the proportion in which they equally limit growth (Fig. 2.6A, B, D). This is the proportion that leads to the corner of the fitness isocline. Thus, optimally foraging plants, if they are morphologically or physiologically plastic, should adjust their morphology and physiology so as to be equally limited by all resources (Iwasa and Roughgarden 1984; Bloom, Chapin, and Mooney 1985). How would this affect the shape of resource-dependent growth isoclines?

Consider consumption constraint curves *a* and *b* in Figure 2.6A. For consumption constraint curve *a*, the optimally foraging individual has the fitness of isocline 1. What would happen if there were an increase in the availability of R_1? This increased availability would mean that, at any given rate of consumption of R_2, this individual would be

FIGURE 2.6. (A–E) These figures differ from all other isocline figures in this book because they show the dependence of growth on the amount of resource consumed, not on environmental resource concentration. The thick lines are nutrition isoclines which show the dependence of the per individual or per unit biomass growth rate on the amounts of two resources that have been consumed (per individual or per unit biomass) in a given time period. The thin lines are consumption constraint curves that show the maximal possible amounts of the resources that can be consumed per individual or per unit biomass in a given time period. The point at which the nutrition isocline of the highest fitness (growth rate) touches the consumption constraint curve is the optimal diet. If there is any tradeoff in the consumption constraint curves (A, B, and D), a plant should adjust its foraging so as to consume the resources in the proportion in which it is equally limited by them. (F) Such optimal foraging would cause the resource-dependent growth isocline, the solid curve in this figure, to have a curved corner rather than a right-angle corner. The right-angle corner isocline shown with a broken line illustrates the underlying physiological constraints. Assuming that there is no cost to plasticity, morphological or physiological plasticity would give the curved isocline shown. If plasticity had a cost, isocline *A* would cross isocline *B*.

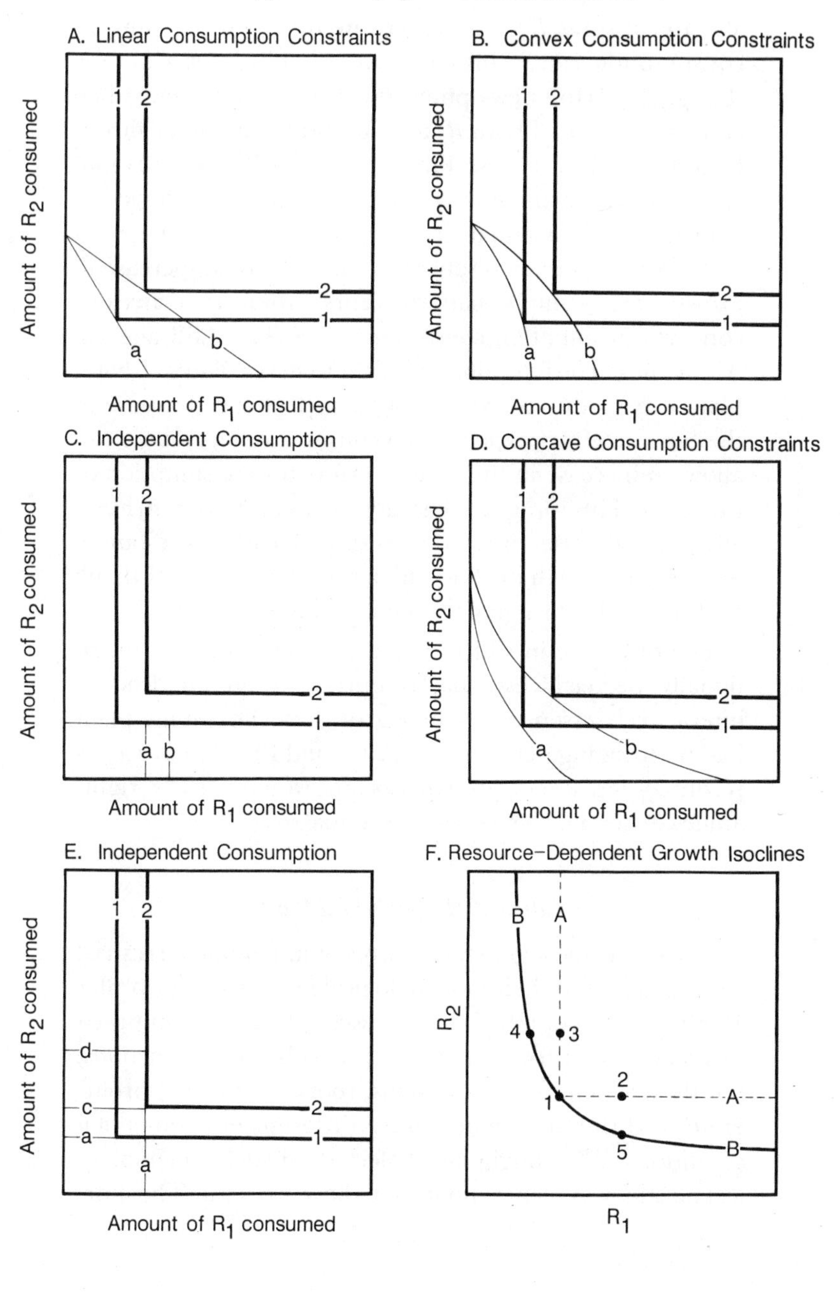
Optimal Foraging for Essential Resources
A. Linear Consumption Constraints
B. Convex Consumption Constraints
C. Independent Consumption
D. Concave Consumption Constraints
E. Independent Consumption
F. Resource-Dependent Growth Isoclines
Amount of R2 consumed
Amount of R1 consumed
R2
R1
1
2
a
b
c
d
A
B
3
4
5

able to consume R_1 at a higher rate than before. Thus, its consumption constraint curve would shift to *b*. Curve *b*, though, leads to a new optimal diet for which the individual consumes more of both R_1 and R_2 and has a higher fitness (growth rate). Thus, even though R_1 and R_2 are nutritionally perfectly essential resources, they behave, ecologically, as interactively-essential resources because of foraging plasticity. Adding either resource would lead to an increase in growth rate. A similar pattern occurs if there are convex or concave consumption constraint curves (Fig. 2.6B and D). All of these consumption constraint curves assume that a plant can modify its consumption such that a decrease in the time or effort expended to consume one resource can allow an increase in the time or effort for consumption of the other. This foraging plasticity means that two nutritionally perfectly essential resources should lead to a resource-dependent growth isocline with a rounded corner. (I thank Joel Brown for bringing this point to my attention.)

The only case in which optimal foraging for two nutritionally perfectly essential resources would not lead to interactively-essential resource growth isoclines is if a plant had no foraging plasticity (Fig. 2.6C and E). In this case, its resource-dependent growth isocline would have a right-angle corner, i.e., be perfectly essential.

Evolution of Morphological Plasticity

One of the most general features of individual terrestrial vascular plants is their morphological plasticity (e.g., Waller 1986). As Harper (1977) has stressed, plants are composed of a set of rather fixed subunits, each with its own functions, but these subunits (leaves, stems, roots, tillers, etc.) proliferate to different extents under different environmental conditions. What might have selected for such morphological plasticity in plants and led it to be so general? The morphology of an individual plant is a physical embodiment of its foraging effort. Roots function in nutrient acquisition. All else being equal, individual plants with greater root mass can acquire more soil resources. Leaves forage for light. Stems also function in light acquisition because stems hold leaves higher, and thus above other leaves where light intensity is greater. Some stems may also be photosynthetic. Because light and a soil resource such as water are nutritionally essential, at any given moment the growth rate of an individual plant should be determined by just one of these. Thus, any plant that could increase its acquisition of the resource that currently limited it could increase its growth rate. Morphologically plastic plants can modify their production of leaves, stems, and roots, and thereby adjust their foraging effort to increase their acquisition of the resource that limits them. They should adjust their morphology so as to be equally limited by all resources (Iwasa and Roughgarden 1984; Bloom, Chapin, and Mooney 1985). A plant that had no plasticity would have a resource-dependent growth isocline with a right-angle corner, such as that shown with a broken line in Figure 2.6F. However, an otherwise identical individual that had a plastic morphology would have the isocline shown with a solid line. The morphologically plastic individual would outcompete the individual with the fixed morphology from all habitats except those that led to the point numbered 1. Thus, individuals that foraged optimally for two essential resources would adjust their physiology and morphology so as to be equally limited by all resources. These individuals would competitively displace individuals that did not optimally adjust their physiology and morphology, assuming that plasticity does not have a cost. The two isoclines in Figure 2.6F may thus explain the ubiquity of morphological plasticity in plants.

CHAPTER TWO

INTERSPECIFIC COMPETITION FOR TWO ESSENTIAL RESOURCES

Because terrestrial plants are morphologically and physiologically plastic (Harper 1977; Waller 1986; Bloom, Chapin, and Mooney 1985), I will assume for the remainder of this book that resource-dependent growth isoclines for above-ground and below-ground resources, such as light and a soil nutrient, are interactively-essential, i.e., have a rounded corner. Resource-dependent growth isoclines are one of the pieces of information needed to predict the outcome of interspecific competition. For an equilibrium to occur, the growth of each species must equal its loss. This occurs for any point on the zero net growth isocline of a species. In addition, equilibrium requires that resource concentrations be constant, i.e., that the rate of consumption of each resource balance the rate of supply. Thus, the actual point on the isocline that will be the equilibrium point in a particular habitat depends on the dynamics of resource supply and consumption.

Resource Consumption

The resource consumption rates of each species can be represented as a vector (Fig. 2.7A). A given species will consume a particular amount of each resource per unit biomass per unit time. If c_1 is the amount of resource 1 (R_1) consumed per unit biomass per unit time, and c_2 is the amount of R_2 consumed per unit biomass per unit time, the total rate of consumption of both resources can be represented as the vector, **C**, which is composed of these two elements.

As illustrated in Figure 2.6, two nutritionally perfectly essential resources should be consumed in the proportion in which they equally limit growth. Excess consumption of a non-limiting resource is of no selective advantage, except as a means of interference competition or in a fluctuating environment. Even in a fluctuating environment, the long-term average rates of resource consumption should be close

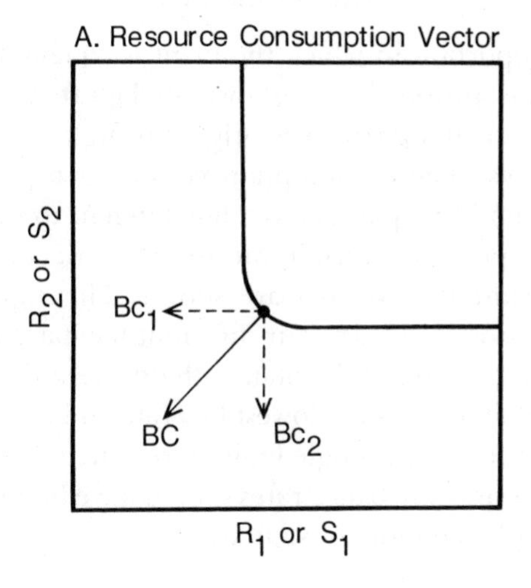

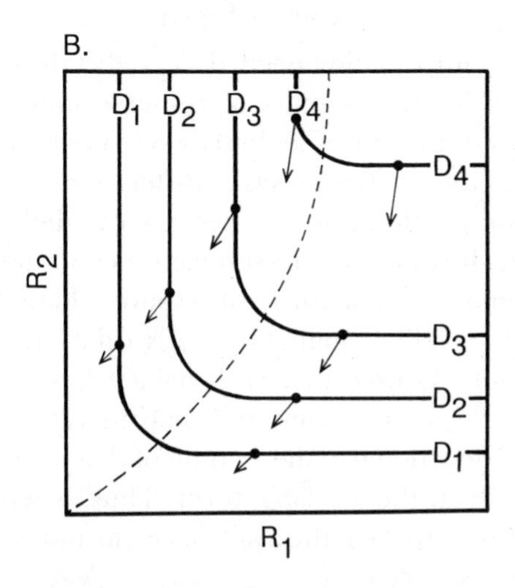

FIGURE 2.7. (A) The solid curve is a resource-dependent growth isocline. The arrows are resource consumption vectors. The consumption vector, $B\mathbf{C}$, has two components, c_1, the consumption rate of R_1 per unit biomass and c_2, that for R_2. These are multiplied by total plant biomass, B, to get total uptake rates. (B) Resource-dependent growth isoclines (solid curve) associated with four different loss rates, D_1 to D_4. The dotted line gives the optimal ratio in which R_1 and R_2 should be consumed by this species at these different loss rates. The optimal ratio is the ratio at which an individual is equally limited by both resources. The arrows are resource consumption vectors.

to the proportion in which the plant is equally limited by them. This proportion is shown in Figure 2.7B by the broken line coming from the origin through the corners of the isoclines. The consumption vector for a point on an isocline should be parallel to a line tangent to this broken line at the point at which it intersects that isocline. Several such consumption vectors are shown. This figure shows isoclines associated with four different habitats, labeled D_1 to D_4. These represent habitats with different death or loss rates, with D_1 having the lowest loss rate and D_4 having the highest. Just as for a single limiting resource, habitats with higher loss or disturbance rates should have higher equilibrial levels of all limiting resources.

Resource Supply

The other information needed to predict the outcome of competition is the dynamics of resource supply. For any actual habitat, this must be directly observed. However, to illustrate, in theory, the effects of different resource supply rates on competition it is only necessary to define a simple rule by which resources are supplied. Let S_1 and S_2 be the maximal amounts of all forms of resources 1 and 2 in a particular habitat. The point (S_1, S_2) is called the resource supply point. Assume that each habitat has a particular, fixed resource supply point and that the rate of resource supply is proportional to the amount of that resource that is not currently in the available form. Thus, where R_j is the amount of resource j in the available form and S_j is the total amount of resource j,

$$dR_j/dt = a(S_j - R_j),$$

where a is a rate constant. This equation defines the rate of supply of resource 1 and of resource 2 in a habitat with a particular supply point (S_1, S_2) and environmental availabilities of the resources (R_1, R_2). These rates of supply can be illustrated as vectors. The rate of supply of resource 1 is

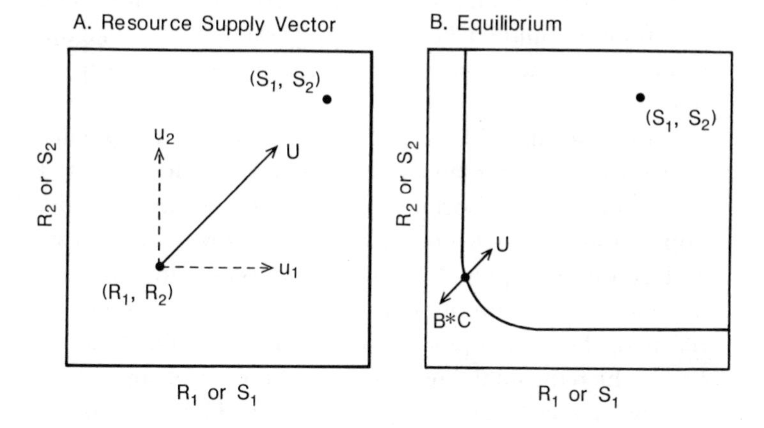

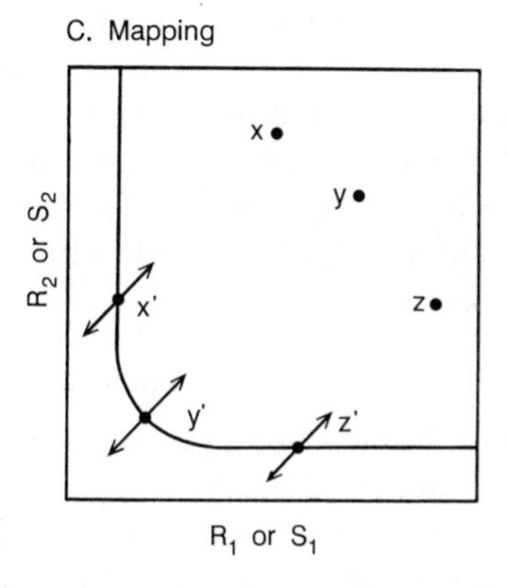

FIGURE 2.8. (A) The resource supply vector, **U**, shows the rates of supply of available R_1 and R_2 for a habitat with a resource supply point of (S_1, S_2). (B) For equilibrium to occur, the total rates of resource consumption, shown by the vector $B^*\mathbf{C}$, must balance total supply, shown by **U**. B^* is total plant biomass at equilibrium. (C) Resource supply vectors and consumption vectors allow each habitat (*x*, *y*, or *z*; where *x*, *y*, and *z* represent different resource supply points) with its particular resource supply rates to be mapped into its equilibrium (x', y', z') on a resource-dependent growth isocline.

shown as the vector $\mathbf{u}_1$ and the rate of supply resource 2 is shown as the vector $\mathbf{u}_2$ in Figure 2.8A. The total rate of supply of both resources, **U**, is the sum of these vectors. The equations defining resource supply rates mean that the resource supply vector will always point toward the supply point.

For the isocline of Figure 2.8B, the resource supply point shown would lead to an equilibrium at the point indicated by the dot on the isocline. At that point on the isocline, population density would be constant (growth equals loss) and resource supply, **U**, would be opposite in direction to resource consumption, **C**. An equilibrium would be reached when the population size (biomass) attained a level, B^*, at which the total rate of consumption by the population exactly balanced the total rate of supply. The total rate of consumption is represented by $\mathbf{C} \cdot B^*$, because **C** is the consumption vector per unit biomass. Thus, at equilibrium, $\mathbf{C} \cdot B^* + \mathbf{U} = \mathbf{0}$. Each supply point has associated with it one particular point on the growth isocline and one particular equilibrium biomass of the consumer species, as illustrated in Figure 2.8C. For a species to survive in a particular habitat (i.e., to have its B^* be greater than zero), it is only necessary that the supply point be "outside" (further from the origin than) its isocline.

Interspecific Competition

There are four distinct cases of two-species competition for two limiting resources (Fig. 2.9). If the isocline of species A is always inside that of species B, species A has a lower requirement for both R_1 and R_2 than does species B, and species A will competitively displace species B from all habitats (i.e., supply points) for which both can survive (Fig. 2.9A). If the isocline of species B is inside that species A (Fig. 2.9B), species B will displace species A from all habitats. It is only if the isoclines cross that there is a two-species equilibrium point. This point is shown in Figure 2.9C as a dot at

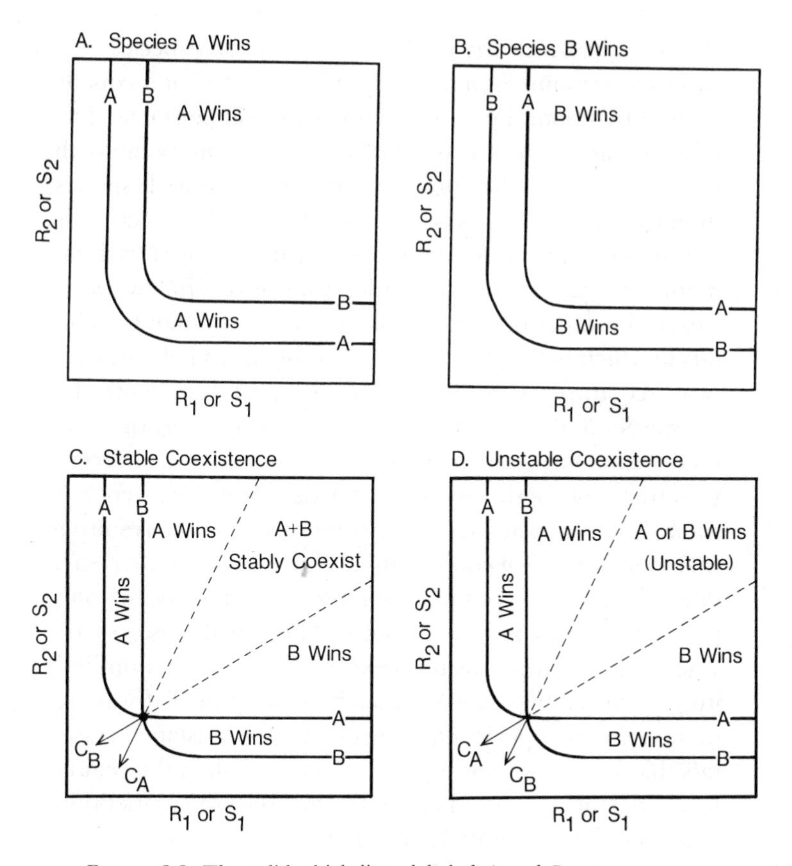

FIGURE 2.9. The solid, thick lines labeled *A* and *B* are resource-dependent zero net growth isoclines for species *A* and *B*. The positions of these isoclines and of the consumption vectors determine the equilibrial outcome of competition in each habitat. (A) Species *A* is a superior competitor for both resources in all habitats (resource supply points) in which either species can survive, and it displaces *B*. (B) Species *B* is a superior competitor for both resources, and displaces *A*. (C) The isoclines cross at a two-species equilibrium point. In combination with the resource consumption vectors of the two species, these isoclines determine the habitat conditions for which species *A* wins, both species coexist, or species *B* wins. Each habitat is defined by its resource supply point, (S_1, S_2). The labeled regions show the outcome of competition expected for supply points that fall in each region. For this case, the equilibrium point is stable. (D) Here, the consumption vectors are reversed. This causes the two-species equilibrium point to be unstable. Either *A* or *B* wins in this region, with the winner determined by initial conditions.

the point at which the isoclines cross. If each species consumes the resources in the proportion in which it is equally limited by them, the equilibrium point will be stable. The isoclines and lines through the equilibrium point with slopes parallel to the consumption vectors of each species define the habitat types (supply points) in which species *A* is dominant, both species stably coexist, and species *B* is dominant. The qualitative explanation for this is as follows. Species *A*, the superior competitor for R_1, is dominant in habitats in which both species are limited by R_1. In habitats with resource supply points for which both species are limited by R_2, species *B*, the superior competitor for R_2, wins. In intermediate habitats, each species is more limited by a different resource, and both can stably coexist. The coexistence is stable because each species consumes more of the resource that more limits it at equilibrium. Expressed in another way, intraspecific competition is stronger than interspecific competition, and coexistence results. The fourth case occurs when each species consumes excess amounts of a non-limiting resource (Fig. 2.9D). If such "non-optimal" foraging occurs, the two-species equilibrium point is unstable. Habitats that have resource supply points that fall in the region labeled "*A* or *B* wins" will have the outcome of competition determined by the initial conditions.

One of the potential advantages of the isocline approach to competition is that it may be able to predict the outcome of resource competition without the need to know all the interrelations of nutrition, habitat structure, and foraging dynamics that determine isocline shape and optimal consumption. Although I have discussed some likely shapes of isoclines and some likely patterns of resource consumption, each of these is best obtained experimentally, not theoretically. Resource-dependent growth isoclines and consumption characteristics can be determined using experimental monospecific stands that are allowed to go to equilibrium and attain stationary size distributions. If such monospecific stands are established on soils that differ in their rate of supply of the limiting soil resource, and if light becomes limiting as plant biomass increases, the isocline would be defined by the level to which light at the soil surface and the limiting soil resource were reduced in each stand. For very long-lived species for which such experiments may be impractical, such as canopy trees, comparisons of available soil nutrient levels and light penetration to the soil surface for a variety of fairly monospecific stands might prove useful in estimating isocline shape and position. Indeed, just such data could probably be collected in monospecific stands of timber and pulp species that were planted many years ago on soils of different site indices. Analyses of nutrient uptake rates and of light interception could give the necessary consumption vectors, or, preferably, the method of analysis presented in Figure 3.11 could be used.

For any given habitat that contained the species whose isoclines were estimated, this approach would suggest that it would just be necessary to determine the rates of resource supply to predict what the equilibrial vegetational composition should be. Such predictions would be based on the assumptions that resource supply rates were constant, that herbivory, disturbance, and other loss rates were constant and identical to those in the habitats used to estimate the resource-dependent growth isocline of each species, and that light intensity at the soil surface, rather than at other positions in the vertical light gradient, was most critical. To determine which species would be predicted to be dominant or which pair would be expected to coexist, it is just necessary to find the point on the isoclines for which consumption vectors would be opposite in direction to supply vectors. For the case of Figure 2.9C, if the resource supply vector had a slope that was intermediate between that of consumption vectors of species A and B, the species should coexist. If the slope of the supply vector did not fall between those of the two species, either species A or B should win.

Species A should win if the resource supply vector, when at the two-species equilibrium point, "pointed" toward the region in which it should win, and species B should win if it "pointed" toward the region in which B should win. This approximation may prove to be useful in many field situations. An alternative approach, which may be preferable when light is a limiting resource, is presented in Figure 3.11.

Multispecies Competition

What pattern does the isocline approach predict for multispecies competition? Let us consider five species competing for a soil resource and light (Fig. 2.10). The isoclines for these species show that they are differentiated in their competitive abilities for these resources. Each species is a superior competitor for a particular proportion (relative availability) of the resources. Each species consumes the resources in the proportion for which it is equally limited by them. These species should become separated along a gradient from habitats with low rates of supply of the soil resource to habitats with high supply rates (Fig. 2.10). At equilibrium, resource levels should be reduced down to points on the isoclines. This means that there should be an inverse relationship between the level of the limiting soil resource and light intensity at the soil surface once equilibrium is reached. Habitats with resource-poor soils (low supply rates) should have lower nutrient levels (concentrations), lower plant biomass, and thus higher penetration of light to the soil surface than habitats with higher supply rates of the limiting soil resource (Fig. 2.10). Each species should attain its maximal biomass in a habitat with a soil

FIGURE 2.10. (A) The lines labeled A–E are zero net growth isoclines for species A–E. The four numbered dots are two-species equilibrium points. The types of habitats (resource supply points) in which various species will be dominant or coexist are shown, based on the assumption of optimal foraging for the resources. (B) These five species should be separated along a soil nutrient gradient, at equi-

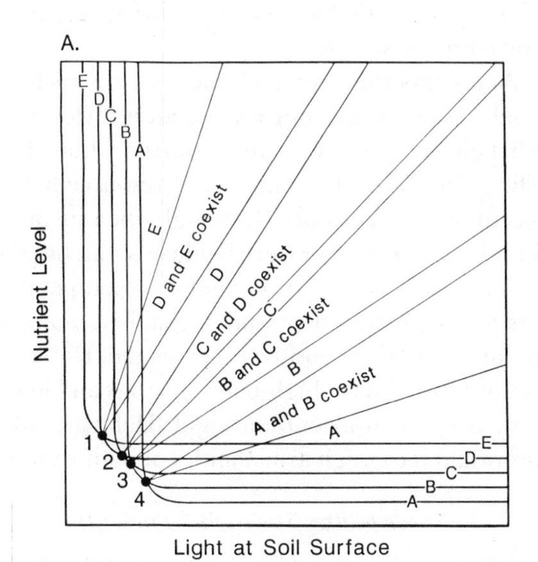

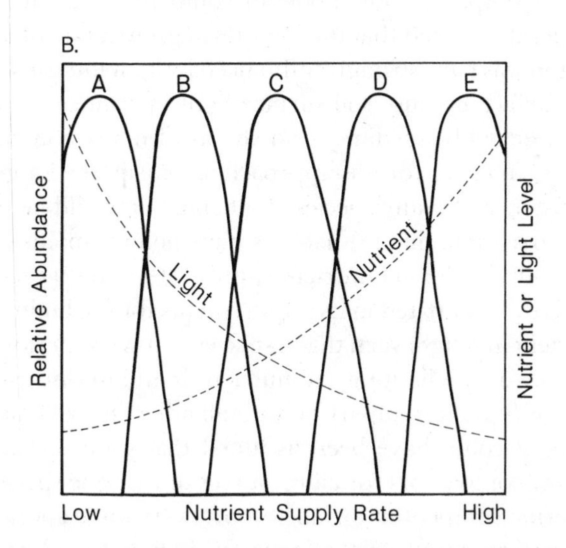

librium, as shown. Habitats with low rates of nutrient supply should be dominated by species *A* and have low nutrient levels but high penetration of light to the soil surface. Habitats with high rates of nutrient supply should be dominated by species *E* and have high nutrient levels but low penetration of light to the soil surface.

resource supply rate that leads to equilibrial resource levels at the corner of its isocline.

For habitats experiencing the same loss rate, each habitat should fall at some point along a gradient from low nutrient but high light availability to high nutrient but low light availability. Qualitatively similar soil-resource:light gradients occur at each loss rate. If, though, the rate of supply of the limiting soil resource were held constant, but the loss rate were allowed to vary, this model of competition for nutrients and light would predict a gradient along which nutrients and light were positively correlated. Habitats with low loss rates would have high plant biomass and low availabilities of both the nutrient and light. Habitats with high loss rates would have high availabilities of both resources.

Supercompetitive Species and Tradeoffs

This simple isocline model of competition for nutrients and light assumed that the long-term growth rate of a population was most strongly influenced by light intensity at the soil surface because soil-surface light intensities are those experienced by seedlings and shoots of newly establishing plants. In order for this approach to explain the long-term persistence of many species in a habitat, it would be necessary to assume that (1) habitats have point-to-point spatial variability in the soil nutrient supply rate, and (2) each species is differentiated in its requirements for the limiting soil nutrient and light such that a species that is a superior competitor for the limiting soil nutrient is an inferior competitor for light, in comparison with all other species. Alternatively, it could have been assumed that some species are poorer competitors for all resources or that one species is a superior competitor for all resources. If some species are poorer competitors for all resources, they should be competitively displaced from all habitats, independent of the resource supply rate. As such, they would be unable to persist even in a spatially heterogeneous habitat. The species persisting in the habitats would be those that were superior competitors for particular resource supply rates, i.e., that had tradeoffs as illustrated in Figure 2.10.

If one species were a superior competitor for all resources compared to the other species, that "supercompetitive" species (Tilman 1982, 1986a) would displace all other species from all habitats, and there would be no differentiation among the species. However, as soon as such a supercompetitive species arose, selection would favor those individuals within that species that became specialized on the particular habitat type in which they lived. This would lead to the evolution either of a new supercompetitive species or of numerous species (or genotypes), each specialized on a particular resource ratio, just as is assumed in Figure 2.10. The specialization would come from each individual giving up some of its ability to use the resource or resources that did not limit it in its habitat in order to increase its ability to acquire and/or more efficiently use the resource or resources that did limit it (Tilman 1982). Thus, of all the possible alternatives for the isoclines of several competing species, it seems likely that coexisting species will be seen to have tradeoffs in their competitive abilities for the limiting resources, assuming, of course, that other biotic or abiotic constraints were not explaining their coexistence. There are several times in this book when I discuss tradeoffs in the abilities of individuals or species to compete for resources. For each of these cases, the objection could be raised that there is no logical foundation for assuming a tradeoff. In every such case, though, it is possible to invoke the appearance of a "supercompetitive" species, and show that selection within such a species would favor an individual with morphological, physiological, or life history traits that increased its competitive ability in the habitat in which it lived, even though these traits decreased its competitive ability in other habitats. Instead of raising this point repeatedly throughout the book, I ask the reader to explore the logic above before objecting to an assumed competitive tradeoff. Ultimately, as discussed in the following chapters, we should be able to understand the underlying morphological, physiological and biochemical processes that cause there to be tradeoffs and thus that have allowed the evolution of the diversity of species we have on earth.

There are many reasons to object to the simplifications made in this approach to plant competition for nutrients and light. One of the least realistic aspects of the theory is the assumption that light at the soil surface is more important than light at other distances above the soil surface. Multicellular, sessile plants live in a vertical light gradient. Light intensity is generally lowest at the soil surface, and it increases, up to the level of full sunlight, at greater distances above the soil surface. Thus, individual plants experience different light intensities as they grow, depending on the heights of their leaves relative to the heights of other plants. In addition, different plant species have different heights at maturity. Further, within a given species, individual plants can be plastic in their morphology, physiology, and life history characteristics. Chapters 3 and 4 present a morphologically explicit model of plant growth and competition. Although more complex, and thus able to address a wider variety of questions (evolution of seed size, of maximal height, of allocation patterns, of maximal growth rates, etc.), this approach can often be approximated by the broad, general theory presented in this chapter.

SUMMARY

This chapter has presented a simple, equilibrium approach to plant competition for limiting resources. Many of the resources required by plants are nutritionally essential. A graphical theory of optimal foraging for essential resources predicts that such resources should be consumed in the proportion in which they equally limit the growth of an individual. Because plants are able to adjust their morphology and thus their foraging effort so as to decrease their consumption of a non-limiting essential resource and thereby increase their consumption of a limiting resource, the resource-dependent growth isoclines of plants should have a rounded corner, not a right-angle corner. As long as there is no cost to such adjustments in foraging effort, an individual plant that is morphologically plastic will be a superior competitor to a plant that is not. The prevalence of morphological plasticity in plants thus agrees with a simple theory of optimal foraging for nutritionally essential resources.

This simple theory also shows that two factors strongly influence the equilibrial pattern of resource availability in terrestrial habitats: the loss rate (from disturbance, herbivory, and mortality) and the rate of supply of the limiting soil resource. Because plant growth rates increase with resource concentrations, habitats with higher loss rates must necessarily have higher levels of all limiting resources, at equilibrium, than habitats with lower loss rates. Thus, one major environmental gradient along which plants may have differentiated is the gradient from habitats with low loss or disturbance rates and low levels of both soil resources and light to those with high loss or disturbance rates and high resource levels. A second major gradient is a productivity gradient from habitats with low supply rates of limiting soil resources to those with high supply rates. Habitats with low soil-resource supply rates should have low nutrient levels, low plant biomass, and high penetration of light to the soil surface, whereas those with high supply rates should have high nutrient availability, high plant biomass, and low penetration of light to the surface. These may be two of the major gradients that plants have experienced during their evolutionary history. If these are major gradients, much of the morphology, physiology, and life history patterns that we see in terrestrial plants could be explained as adaptations to competition along these two axes. This question is explored in Chapters 3 and 4.

CHAPTER THREE

Mechanisms of Competition for Nutrients and Light

Although the isocline theory presented in Chapter 2 may summarize the process of resource competition among terrestrial vascular plants, it does not include the explicit morphological, physiological, and life history mechanisms whereby plants compete for light and soil resources. Because it does not include these factors, it cannot be used to determine which plant morphologies, life histories, and physiologies should be favored in habitats with particular rates of supply of limiting soil resources and particular loss rates. In this chapter, I present a mechanistic model of competition among size-structured plant populations, and explore some of its equilibrium and dynamic features. In the following chapter, I use this model to determine how plant traits should be influenced by loss rates and by resource supply rates. This model assumes that the pattern of growth of an individual plant is determined by its pattern of allocation of current photosynthate to the production of roots, leaves, stems, or seeds. Before discussing the model, I must present a general, simple, but critically important relation between allocation patterns and maximal growth rates.

ALLOCATION AND MAXIMAL GROWTH RATES

Until about 400 million years ago, the only terrestrial photosynthetic plants were cyanobacteria and green algae. From that time to the present there has been an amazing proliferation of perhaps a million or more species of multicellular, terrestrial plants, all of which were derived from some ancestral algae (Stebbins and Hill 1980). These terrestrial species show a wide array of differing morphologies, physiologies, growth rates, and life histories. Yet, in some sense, the difference between a soil alga and a giant sequoia is mainly a difference in morphology, caused by the pattern of allocation of photosynthetic production. Once a single-celled alga has acquired sufficient resources to approximately double its size, it divides into two individuals. This doubling can occur twice a day under ideal conditions. A redwood seedling, in contrast, allocates much of its potential growth to roots, stems, and needles in such a way that its total size increases by many orders of magnitude over a period of hundreds of years. Eventually, it allocates some of its production to seed. Although plant species differ in many ecologically important biochemical and physiological characteristics, the most striking feature of the evolution of vascular plants from single-celled algae is the different patterns of allocation to plant structures such as roots, stems, leaves, and seeds. Schopf et al. (1983, p. 382) state that by about 1.6 billion years ago, the "early, metabolically and biochemically based successional development of the Earth's ecosystem had been completed, supplanted during the late Precambrian by a new, successful, morphologically based mode of evolutionary advance." The most parsimonious explanation for the intraspecific and interspecific morphological and life history patterns that we observe today is that the same selective forces that led to the evolution of these patterns maintain these patterns today.

Maximal Growth Rates

In trying to explain the life history patterns that are seen in modern vascular plants, it is worthwhile to remember their ancestry, and a major contrast between algae and vascular plants. Algae have very high maximal growth rates, being capable of doubling their biomass up to twice a day (RGR_{max} = 1.4 day^{-1}; where RGR_{max} is the maximal long-term *r*elative *g*rowth *r*ate for an ideal, unlimited habitat, calculated as $dB/dt \cdot 1/B$, where B is plant biomass). The most rapidly growing terrestrial vascular plant seedling reported by Grime and Hunt (1975), the annual *Poa annua*, has a maximal rate of vegetative growth that is only 1/4 of this (RGR_{max} = 0.38 day^{-1}). The most slowly growing plant seedlings reported by Grime and Hunt, spruce (*Picea sitchensis*) tree seedlings, have a maximal growth rate of RGR_{max} = 0.03 day^{-1}, an order of magnitude slower than *Poa annua*. The early history of vascular plant evolution was a progression from single-celled algae (Stebbins and Hill 1980) to multicellular plants of short stature, to progressively taller, branched plants. How is it that slowly growing forms evolved from rapidly growing forms? What selective advantage gained by individual plants could have more than compensated for such dramatic decreases in their maximal rates of vegetative growth? To answer this question, it is necessary to consider the forces that determine the maximal growth rates of plants.

Consider a simple model of plant growth, as illustrated in Figure 3.1. In theory, a plant can have almost any pattern of allocation to photosynthetic tissues, structural and nutrient transport tissues, nutrient absorptive tissues, and reproductive tissues. For vascular plants, these tissues would correspond, approximately, to leaves, stems, roots, and seeds. How might allocation to these different structures influence the maximal rate of vegetative growth of a plant?

The maximal vegetative growth rate is the greatest growth rate that a plant can attain in an ideal habitat, i.e., in a habitat in which it is not limited by any resources. In such a habitat, growth is an exponential process. It is an exponential process because of the continual compounding of growth through reinvestment of production. If all current

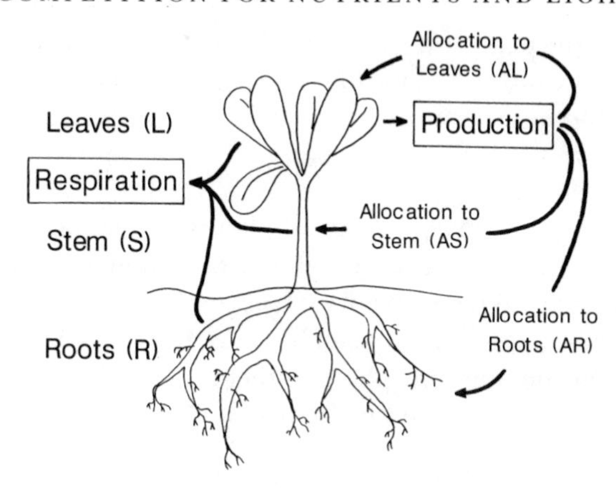

FIGURE 3.1. A diagramatic representation of growth in a plant that can allocate its production to additional leaf, stem, or root biomass. Equation 3.4 models growth of such a plant in a nutrient-saturated environment. The model ALLOCATE, presented later in this chapter and in the Appendix, models the growth of plants in resource-limited habitats.

growth were allocated to the production of new photosynthetic tissues (which is approximately the case for a single celled alga living in a high nutrient and high light habitat) a plant with a given maximal rate of photosynthesis would be able to attain its most rapid rate of vegetative growth (Monsi 1968). To the extent that some of its photosynthesis were used for purposes that did not directly lead to increased photosynthetic tissues, the maximal growth rate would be decreased. Thus, plants that allocate some of their growth to stems or roots or other non-photosynthetic structures necessarily have lower maximal rates of vegetative growth than do plants that allocate all growth to additional photosynthetic tissues.

This may be illustrated by considering the following differential equations, which are a simple mathematical embodiment of the flow chart of Figure 3.1. Let's first consider a case in which a plant allocates all of its potential growth to photosynthetic structures. Let B be the total biomass of the plant and let L be the biomass of photosynthetic structures (leaves). The rate of change of plant biomass, dB/dt, is thus

$$dB/dt = LP_m - rB \quad (3.1)$$

where P_m is the maximal rate of photosynthesis per unit of photosynthetic tissue under optimal conditions (e.g., grams of carbon fixed per gram of carbon per day) and r is the respiratory rate per unit of plant biomass. Here $B = L$ because the plant contains only photosynthetic tissues. Because $B = L$, this equation can be rewritten as

$$RGR_{max} = dB/Bdt = P_m - r \quad (3.2)$$

where dB/Bdt is the maximal rate of vegetative growth of this plant, with units of $time^{-1}$. This equation simply states that the maximal growth rate is equal to the difference between the resource-saturated photosynthetic rate and the respiration rate. Given a plant with this P_m and r, how would its maximal rate of vegetative growth be affected if it allocated some of its production to such non-photosynthetic structures as stems and roots?

Because we are interested in maximal growth rates, we will consider again cases in which the plant is living in an unlimited habitat. Let S be the amount of total plant biomass in stems and R be the amount in roots. This gives $B = R + S + L$. Then, of the total plant biomass, B, only an amount equal to L, where $L = B - S - R$, is photosynthetic tissue. Substituting $B - S - R$ for L in Equation 3.1 gives

$$dB/dt = (B - S - R)P_m - rB, \quad (3.3)$$

and

$$RGR_{max} = dB/Bdt = \mathbf{P}_m[1 - (R + S)/B] - r, \quad (3.4)$$

or

$$RGR_{max} = (L/B)P_m - r. \quad (3.5)$$

What does Equation 3.4 imply? The term $(R+S)/B$ is the proportion of the plant's total biomass that is allocated to roots and stems, i.e., to non-photosynthetic tissues. Thus, the maximal rate of vegetative growth of this plant under ideal environmental conditions is decreased by an amount equal to the proportion of total plant biomass allocated to non-photosynthetic tissues. Maximal growth rate is determined by the proportion of total production that is allocated to leaves (Eq. 3.5). A simple but major conclusion to be drawn from Equation 3.4 is that any allocation to any structures other than photosynthetic structures necessarily leads to a decrease in the maximal, resource-saturated growth rate of an individual plant. Is this prediction supported by experimental evidence?

Grime and Hunt (1975) measured the maximum relative growth rates of seedlings of 132 species of flowering plants of the Sheffield, England, region. They grew each species for five weeks with a single individual per container in nutrient-rich sand culture at high light intensity. The results of their experiments are summarized in Figure 3.2, for which I used the RGR_{max} values in conjunction with the description of the growth form (grass, forb, legume, or woody plant) and life history (annual or perennial) provided by Clapham, Tutin, and Warburg (1962). As a group, seedlings of annual plants, which do not allocate resources to perennating structures (woody stems, woody roots, energy and nutrient stores, buds) had significantly higher maximal growth rates than herbaceous perennials. Herbaceous perennials had higher maximal growth rates than woody perennials.

This progression from annuals to herbaceous perennials to woody perennials represents a gradient from low to high allocation of production to non-photosynthetic structures. The most slowly growing group, woody perennials, allocate much of their growth to the production of the woody roots and stems, to perennating tissues, and to energy stores that are used for growth early in the season. Although differences in the loss and respiration rates of roots, leaves, and

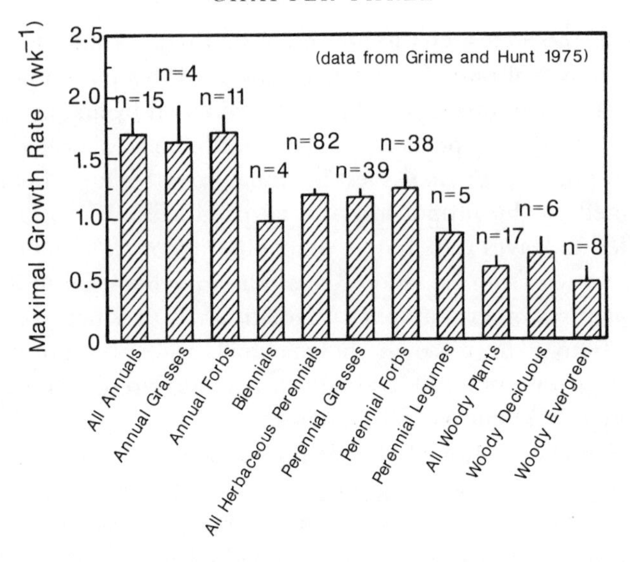

FIGURE 3.2. Dependence of average maximal growth rates (bar) and standard errors (line) on plant life form and life history. Maximal growth rates calculated as dB/Bdt of seedlings grown under high light and high nutrient conditions. Figure prepared using data from Grime and Hunt (1975). Note that some data are shown twice because both groups (e.g., all annuals) and subgroups (e.g., annual grasses or annual forbs) are shown.

stems mean that allocation patterns can be quite different from the proportion of biomass in these structures, woody plants often have 70%-90% of their biomass in stem, 9%-25% in roots, and as little as 1%-6% in leaves (Kira 1964; Whittaker 1975). Plants with such morphologies should have low relative growth rates. Interestingly, Tamm and Aronsson (1982) found that the increase in the growth rate of spruce trees following fertilization corresponded with increased allocation to needles and decreased allocation to fine roots. The most rapidly growing group, annuals, have a high proportion (50%-70%) of their production in leaves as seedlings (Harper and Ogden 1970; Howrath and Williams 1972). Thus, the data presented by Grime and Hunt are qualitatively consistent with the predictions of Equation

3.5. A direct quantitative test of the predictions of Equation 3.5 would require more information on allocation patterns in these species than is currently available. Grime and Hunt (1975, p. 412) did note, though, that the slower growing species within a group often had a high allocation to roots, including development of long, swollen, tap root systems. Further, perennial legumes had significantly lower maximal growth rates than other perennial herbaceous plants (Fig. 3.2). To maintain their symbiotic association with nitrogen-fixing bacteria, legumes must provide these bacteria with photosynthate. This increased allocation of photosynthate to a function other than production of more photosynthetic tissues may explain the lower maximal growth rates of legumes and their absence during secondary succession on rich soils (Chapter 7).

This analysis of Grime and Hunt's data, as well as earlier work by Jarvis and Jarvis (1964), Monsi (1968), and Loach (1970), demonstrates that allocation of photosynthate to structures other than leaves generally reduces the maximal, resource-saturated growth rates of plants. Such differences in maximal growth rates are of great ecological importance, because the maximal growth rates of plants, as discussed in Chapter 6, are a major determinant of their short-term dynamic responses to any change in their environment, be it an experimentally imposed manipulation or some biotic or abiotic disturbance. Clearly, Equation 3.5 is an approximation to the possible dependence of maximal growth rates on plant morphology. It assumes that leaves, roots, and stems have equal respiration rates per unit of biomass, and that they have similar turnover rates. Neither of these is a particularly realistic assumption. However, the qualitative prediction of Equation 3.4 is so robust that inclusion of these factors does not change it. For instance, roots and especially stems may have lower respiration rates than leaves. Let's assume an extreme case, in which the respiration rates of leaves is r_L and that of stems and roots is zero.

This would give, for a resource-saturated plant,

$$RGR_{max} = dB/Bdt = L/B(P_m - r_L), \qquad (3.6)$$

which has higher values for RGR_{max} at all proportional allocations to leaves, L/B, than Equation 3.5. Leaves may also have higher turnover rates than stems. Such turnover functions as a cost, like respiration, that reduces the net production per leaf. A comparison of Equation 3.4 with Equation 3.6 (Fig. 3.3) shows that the assumption of equal respiration or turnover rates for all tissues (Equation 3.4) and the assumption of leaves being the only tissue with a respiratory or turnover cost (Equation 3.6) provide likely bounds on the dependence of RGR_{max} on allocation patterns. Because different structures are likely to have different respiratory and turnover costs, a linear relation is unlikely to occur. However, the actual function should be bounded between the two lines of Figure 3.3, giving maximal growth rates that increase with the proportional allocation to leaves. To apply the relations of Figure 3.3 to many different species, in a comparative manner, as I have done above, it is necessary to assume that species-to-species differences in morphology are much greater than species-to-species differences in nutrient- and light-saturated rates of photosynthesis per unit biomass or respiration rates per unit biomass. This assumption is probably a valid first approximation. Terrestrial plants range from soil algae that are functionally almost 100% "leaf" to canopy trees that are less than 1% leaf. Within a given geographic region, the maximal specific rates of photosynthesis under field conditions typically vary by less than tenfold (see papers in Chabot and Mooney 1985). However, after morphology, the next major determinant of maximal growth rates is likely to be maximal rates of photosynthesis. Interestingly, there seems to be more genotypic variation in morphology than in physiology. Thousands of years of selection for highly productive cultivars of various crop plants have had little effect on physiology but a major effect on allocation patterns and mor-

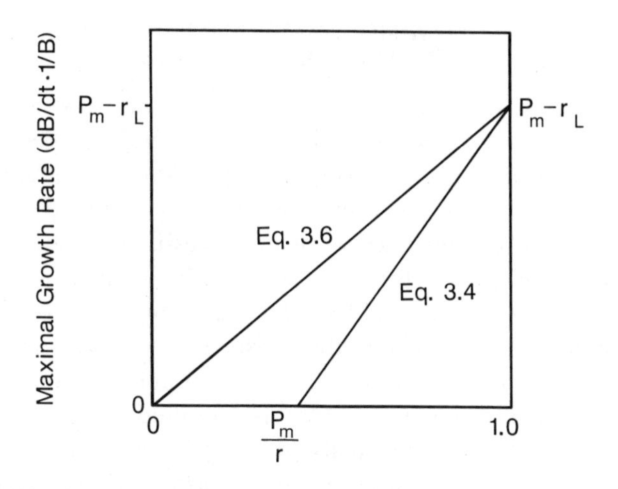

FIGURE 3.3. Maximal growth rate is predicted to increase with the proportion of photosynthetic production that is allocated to leaves. The two curves shown, one from Equation 3.4 and the other from Equation 3.6, represent likely bounds on the dependence of maximal growth rate on leaf allocation. Both curves assume that maximal photosynthetic rates (P_m) and respiration rates (r) are held constant.

phology (Fitter 1986). Gifford and Jenkins (1981) found that allocation to leaf was a stronger determinant of growth rates of crop plants than maximal photosynthetic rates.

All else being equal, plants with greater maximal growth rates should be favored over those with lower maximal growth rates. How is it, then, that slow-growing plants evolved from rapidly-growing plants? In order to explain the evolution of morphologies that result from allocation to structures other than photosynthetic tissues, it is necessary to determine what advantages an individual gains from such structures when it is living in a resource-limited environment. In an unlimited environment, there is no selection favoring multicellularity and the evolution of non-photosynthetic tissues, unless such structures reduce losses or mortality.

CHAPTER THREE

ALLOCATE: A MODEL OF COMPETITION IN SIZE-STRUCTURED PLANTS

The only way to address these questions with any rigor is to construct models that explicitly include critical aspects of plant morphology, physiology, and life history as well as the complexity of the plant environment. Unfortunately, the fascinating complexities of plants necessitate mathematically complex models that are analytically tractable only in extremely special cases. I have developed one such dynamic model, called ALLOCATE, that includes several aspects of complexity that I believe may be of critical importance for plants. The elements included in the model are (1) size structure, with individuals starting as seedlings and undergoing continuous vegetative growth until they attain their maximal height; (2) vegetative growth determined by allocation of photosynthate to roots, leaves, and stems; (3) a vertical light gradient imposed by plant interception of light; (4) nutrient and light dependent photosynthesis; (5) nutrient availability in the soil determined by mineralization rates and rates of nutrient uptake by plants; (6) seed size, which influences both the number of seeds produced by a plant and, along with allocation patterns, the height, root mass, leaf mass, and stem mass of seedlings upon germination; (7) potentially different respiration rates for roots, leaves, and stems; and (8) density-independent loss (specifically mortality) rates. The only mechanism of interaction between individual plants—be they of the same or different species—assumed by the model occurs through resource consumption. Thus, there are no directly density-dependent processes assumed in the model. Rather, each individual influences its own growth and that of other individuals solely through its effect on the vertical light profile and on soil resource availability.

The model does not conceptually distinguish between competition caused by shading and competition caused by nutrient consumption. As Harper (1977) noted, there are differences between these two processes, but these differences are directly included in this model because the height of each plant, and thus the light environment in which it lives, is an explicit part of the model. A major simplification of the model is that there is no horizontal structure to the habitat, just vertical structure. Thus, at any given point in time, all plants experience the same soil nutrient concentration (level) and all plants of a given height experience the same light intensity. As such, this should be considered a model of plant dynamics within a small patch that has a homogeneous soil but lacks light gaps and "neighborhood" competition (sensu Pacala and Silander 1985). This model, as are all models, is a caricature of reality, and, as May (1973, p. 12) noted, has "both the truth and falsity of all caricatures."

One of the purposes in constructing and solving, via simulation, this model has been to seek general patterns that may come from it so as to determine how valid simpler models that ignore such complexity, such as isocline models, may be. Another purpose has been to determine how plant morphology, seed size, and height at maturity should change along major environmental gradients. A third goal has been to determine the implications of plant morphologies and life histories for the dynamics of community changes following natural disturbances or experimentally imposed manipulations. However, simulation models are never elegant. They go well beyond the domain that any respectable mathematician would attempt to deal with in an analytical manner. It is impossible, within the limits of this book and time available to me on a supercomputer, to rigorously explore all the interactions among all variables in this model and map out their full dynamic behavior. These disclaimers aside, though, there are many insights that can be gained from such models. I do not believe that any of the qualitative generalizations offered in the following chapters would be changed in any major way if I modified the actual form of ALLOCATE in any biologically realistic manner. It includes the important aspects of competition among size-structured plants living in a habitat with a homogeneous soil substrate, a single limiting soil resource, a vertical light gradient imposed by the plants themselves, and density and size-independent mortality.

The Appendix following Chapter 9 presents the mathematical details of this model of plant allocation and growth. I will summarize, below, the major assumptions made by this model, and present the logical outcome of these assumptions. In so doing, I want to remind the reader of the purpose of modeling. All models are logical devices that convert a series of assumptions into predictions. Mathematics is merely the tool that we use in this conversion. Assuming that the mathematics is correct, the important features of a model are its assumptions and its predictions. If a model makes predictions that are shown, through either experimentation or observation, to be wrong, then one or more of the underlying assumptions of the model are incorrect. Because of the complexity imposed by having different individuals within a species be different heights, and thus experience different light intensities, I found no way to solve this model analytically, but used simulations to determine the predictions associated with the biological assumptions I made.

As Harper has repeatedly emphasized (Harper 1977), plants compete for both soil nutrients and for light. Because plant height is a major determinant of competitive ability for light, a mechanistic model of plant competition must allow plants to differ in their heights, and thus in the proportion of the total incident light that they can receive. Plants start their life as seedlings or shoots or runners, and grow vegetatively. As their height increases, they capture more light. Similarly, plants with larger root systems can obtain more of a limiting soil resource. ALLOCATE has each individual plant start as a seed (Fig. 3.4). Upon germination, the energy and nutrients stored in a seed are used to produce roots, stems, and leaves. The proportion of seed biomass allocated to roots, stems, and leaves is fixed, independent of environmental conditions, and is the same as the proportion of photosynthate that is allocated to these structures during vegetative growth. Each plant grows vegetatively until it reaches its maximal height. Once a plant reaches its maximal height, the model assumes that all photosynthate that is in excess of respiration is allocated to the production of seed. Such a switch from vegetative growth to reproduction occurs in many, but by no means all, species. It is a useful first approximation for modeling, and was predicted to occur in a model of optimal growth of an annual plant (Iwasa and Roughgarden 1984).

ALLOCATE assumes that all life history stages are subject to density-independent mortality, with mortality being size-independent and resource-independent. This is a useful first approximation because of its simplicity. Under conditions of high nutrient availability, photosynthesis can be light limited, with photosynthesis assumed to be a Michaelis-Menton function of light intensity. Under conditions of high light intensity, photosynthesis can be nutrient limited, with photosynthesis being a Michealis-Menton function of the "relative nutrient availability," which is defined as the concentration of available nutrient in the soil multiplied by the ratio of root mass to leaf mass. Further, it is assumed that the limiting nutrient and light act as perfectly essential resources in controlling the rate of photosynthesis. This means that the rate of photosynthesis is determined by either the level of the limiting nutrient or the light intensity, whichever leads to the lower rate.

The model assumes that the photosynthate produced in a given interval of time will first be used to cover the costs of respiration of the existing tissues during that time interval, and that the remaining photosynthate will be allocated to the production of new roots, stems, and leaves (or to seeds

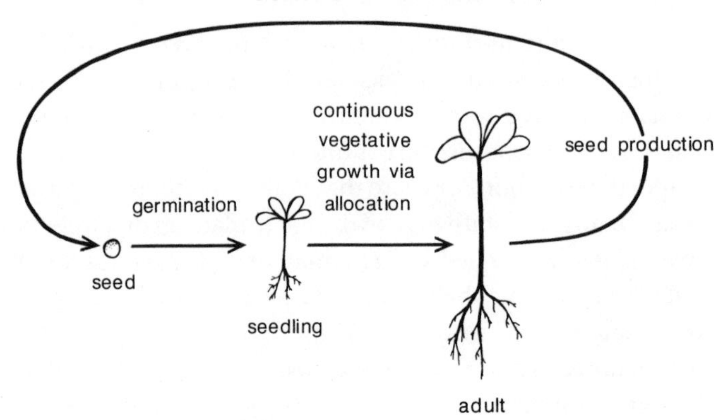

FIGURE 3.4. Life cycle of a plant as modeled by ALLOCATE. Individuals start life as seeds and undergo continuous but resource-dependent vegetative growth until they reach their maximal size (adult). At that point, all photosynthate that exceeds respiration and other losses is allocated to seed production.

if the plant has attained its maximal height). If respiration exceeds photosynthesis, the plant is assumed to become dormant, as has been reported for many plant species (Fitter 1986). Dormant plants are still subject to mortality. The photosynthate allocated to various structures is determined by the proportional allocation to roots (AR), to leaves (AL), and to stems (AS). New allocation to leaves and stems increases leaf and stem biomass and thus changes the light environment for all shorter individuals.

In ALLOCATE, the amount of light an individual receives is determined by the height of its leaves and the vertical distributions of the biomass of all other plants. All leaves are assumed to be born in a monolayer at the top of the stem. Leaf height is determined by stem height. Stem height is assumed to be a function of stem biomass. The function used is based on structural constraints (e.g., McMahon 1973) and allometric relationships (e.g., Kira 1964) that have been reported between plant height and stem biomass, as well as my own unpublished observations of herbaceous plants.

Each tissue type (root, stem, leaves, or seed) has its own nutrient content. Nutrients are taken up in the amounts required to meet the nutrient demand established by new production of roots, stems, leaves, and seeds. This eliminates the possibility of luxury consumption, i.e., nutrient consumption in excess of current need. For each tissue, this is done by multiplying the amount of nutrient required per unit biomass of that tissue by the biomass of newly produced tissue of that type.

The limiting soil nutrient is assumed to be supplied in an "equable" mode (Tilman 1982), i.e., the nutrient supply rate is proportional to the pool size of the potentially mineralizable forms of the limiting soil nutrient. I will call this TN, for "total nutrient." I do not wish to imply that nutrient supply need always be proportional to the total pool of a limiting soil nutrient, for this need not be the case (Pastor et al. 1984). This should serve as a useful first approximation for the dynamics of supply of soil nutrients such as nitrogen or phosphorus because they have an unavailable form that gets converted into an available form. This may also be a useful first approximation for water, because the supply rate of water depends both on rainfall and on the unused portion of the total water-holding capacity of the soil. The lower the level to which plants reduce soil moisture, the greater would be the amount of soil moisture recharge that could occur. This effect would cause soil moisture recharge rates to depend on the precipitation rate in much the same way that nitrogen or phosphorus supply rates would depend on total nitrogen or total phosphorus. As such, TN can be considered a measure of the supply rate of any limiting soil resource, with the supply rate being roughly proportional to TN.

The model operates by allowing each cohort of each species or phenotype to grow simultaneously. As they grow, their consumption reduces the availability of soil nutrients. Light is attenuated by the leaves and stems of the plants according to Beer's Law. The proportion of the incident photosynthetically active radiation that reaches a leaf at a given distance below the top of the canopy is thus a negative exponential function of the total plant biomass at all levels above that leaf. The model is solved numerically using approximately daily time increments. In honor of the climate of Minnesota, a "year" was considered to consist of 140 growing days. During a given time increment, each cohort of each phenotype or species grows according to the light and nutrient environment it experiences. The light and nutrient environment for the next time period are then established by the cumulative effect of their uptake and stem and leaf production. This process is repeated until an equilibrium solution is approached or sustained oscillations are attained. The model does not include any seasonality, but rather assumes continuous growth. Thus, the model assumes that the only mechanism of intraspecific and interspecific interaction is through resource consumption.

The model was written to be quite general. Each phenotype (or species) can be given its own allocation pattern, height at maturity, seed size, maximal photosynthetic rate, light and nutrient half-saturation constants for photosynthesis, and tissue nutrient concentrations. For most of the simulations, I used parameters that were meant to represent grassland species, with nitrogen as the limiting soil nutrient. Thus, the greatest maximum height used was 2.5 m, the minimum possible height was 0.6 cm, maximum growth rates spanned those reported by Grime and Hunt (1975) for herbaceous species, and nutrient-limited growth rates were comparable to those reported by Tilman (1986b). This choice of parameters, though, does not affect the qualitative nature of the patterns predicted by the model.

Growth of a Single Phenotype

Let us first consider the dynamics of growth of a single species in a particular environment. Depending on the parameters chosen, there are two different dynamic responses that occur. First, the population can approach an equilibrium density with many other parameters also going to equilibrium. Second, the population can exhibit sustained oscillations in its total density and in resource levels. Let's first consider a case (a habitat with a very low nutrient supply rate; TN = 50) in which an equilibrium is attained (Fig. 3.5A–C). As the population grew, and total plant biomass increased, the proportion of light penetrating through the vegetation decreased as did the availability of the soil nutrient. After about one year, the population reached its equilibrium total biomass. Light penetration to the soil surface and available soil nutrient also approached constant values. Even though total biomass, light at the soil surface, and available nutrient became constant, the population did not necessarily attain a stable size distribution. For the case of Figure 3.5A–C, for instance, the size structure had sustained oscillations, with "waves" of seedlings growing into adults, dying, and being replaced by a new wave of seedlings. Total biomass, light penetration, and available nutrient levels remained essentially constant, though, because all plants, independent of their size, were nutrient limited on this poor soil. Total biomass was held constant by this uniformity of nutrient limitation.

In the second situation, an equilibrium is never reached. Instead, total population density, the relative abundances of each age class, and nutrient and light availabilities undergo sustained oscillations (Fig. 3.5D–I). From numerous simulations I have performed, such oscillations are common on soils with sufficiently high rates of nutrient supply that some individuals (the shorter ones) are light limited and others (the taller ones) are nutrient limited. Two such situations are shown in Figure 3.5D–I. When a population is nutrient limited, there is little total effect of size structure on the growth of the population. However, once it is light limited, there is a tendency for a cohort of individuals to grow up to adult size, and produce many off-

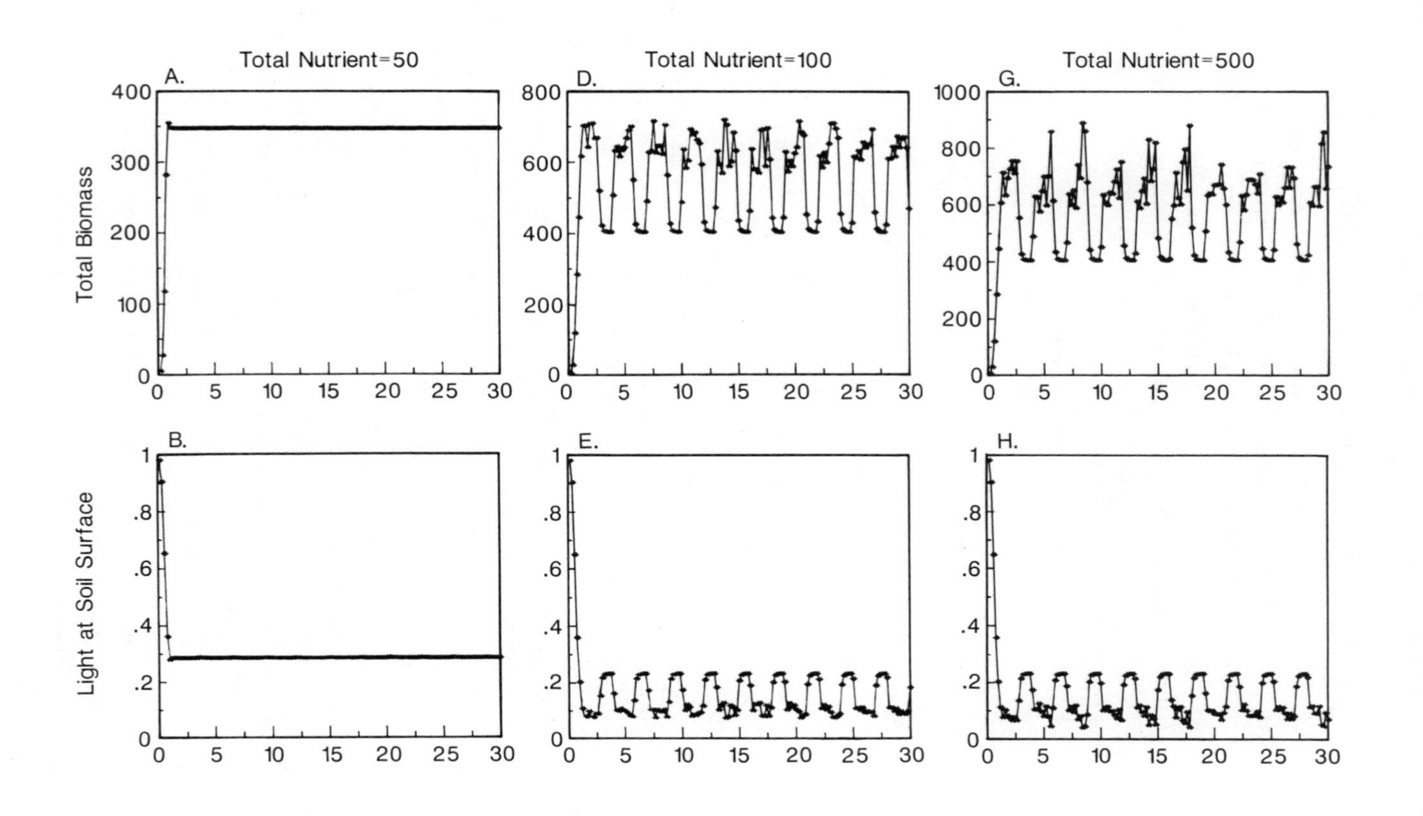

Total Nutrient=50
Total Nutrient=100
Total Nutrient=500
A.
B.
D.
E.
G.
H.
Total Biomass
Light at Soil Surface
0
100
200
300
400
600
800
1000
.2
.4
.6
.8
1
5
10
15
20
25
30

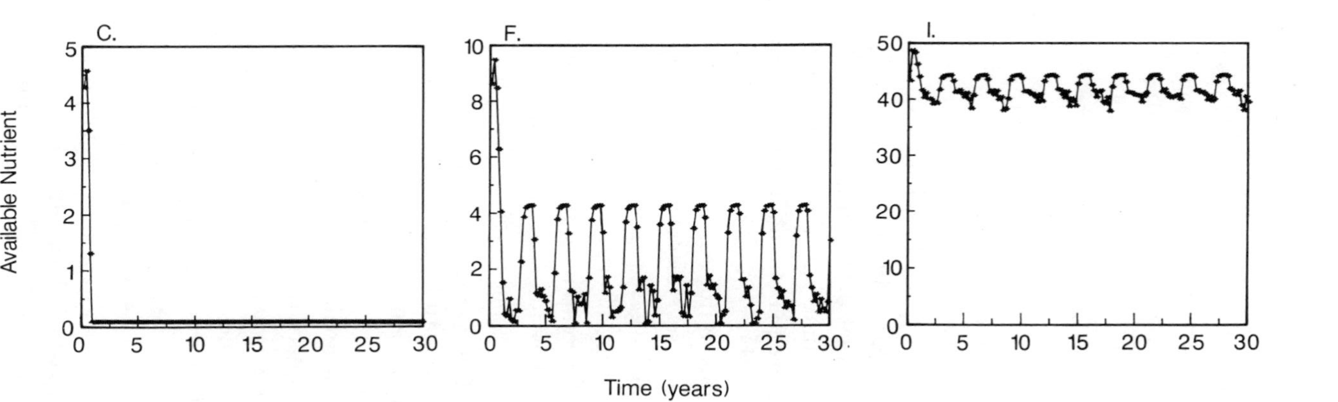

FIGURE 3.5. The dynamics of growth of species *A* in three different habitats. (A, B, and C) In a nutrient-poor habitat (TN = 50) in which all individuals, independent of their size, are nutrient limited, total biomass (*A*), light at the soil surface (*B*) and available nutrient (*C*) rapidly attain and maintain a quasi-equilibrium. (D, E, and F) In a more nutrient-rich habitat, with TN = 100, some individuals are light limited and there are sustained oscillations. (G, H, and I) There are also sustained oscillations in an even more nutrient-rich habitat (TN = 500). TN, which stands for "total nutrient," is a measure of the nutrient supply rate of a habitat. The traits of species *A* are described in the following section titled "Interspecific Competition."

spring. These offspring are inhibited by the shade cast by the canopy individuals. Thus, there is a strong tendency for most seedlings to be suppressed and to grow up to be canopy individuals only when the current canopy individuals start to die. Oscillations are common in many age- or size-structured models of population growth (e.g., Sonleitner 1977), because age or size structure introduces time delays into a density-dependent process. As May (1973) noted, time delays tend to decrease population stability. I should note that, as would be expected, oscillations tend to be greater the greater the maximal growth rate of a species or the longer the time interval over which the model is solved numerically. Longer time scales make the model behave more as a difference equation model than as a differential equation model (May 1973). I chose the time interval for most simulations in this book by trying a wide variety of time intervals for several cases of two-species competition, and using the largest possible time interval that prevented any time-interval-dependent oscillations. However, I did not have the computer resources to assure that this was necessarily the best interval for all other cases. This interval was approximately one iteration per "day."

Figure 3.5 shows the responses of a size-structured plant species, species *A*, in habitats with three particular rates of supply of the limiting soil nutrient. How might this species respond at different points on a productivity gradient? Because of the tendency for oscillations, especially at high rates of nutrient supply, for each nutrient supply rate I calculated the long-term average value of each variable (plant biomass, available soil nutrient, light intensity at various heights) over several oscillations, once sustained oscillations had been obtained (Fig. 3.6). The average population density of species *A* in monospecific stands increases with the rate of nutrient supply, and then asymptotes at high rates of supply. The average level to which the limiting soil nutrient is reduced by species *A* is constant at low rates of nutrient supply, and then increases with the rate of nutrient supply. The amount of light penetrating to any given distance below the top of the canopy has the opposite pattern. For instance, light penetration to the soil surface (Fig. 3.6C) and light penetration to a height 17 cm above the soil surface (Fig. 3.6D) are both high, on average, in habitats with low rates of nutrient supply. In progressively more nutrient rich habitats, an increasingly smaller proportion of incident light penetrates to the soil surface and to a height 17 cm above the soil surface. In even richer habitats, the proportion of incident light reaching the soil surface reaches an asymptote and is constant, independent of the rate of nutrient supply, as is total plant biomass.

Resource-Dependent Growth Isoclines

For each nutrient supply rate, there is an average level to which the limiting soil nutrient is reduced and an average vertical light gradient imposed by the vegetation after this population has grown in monoculture for a long time. For the time being, let us assume that light penetration to the soil surface is indicative of the effect of light on the growth of a species. Then, for each rate of nutrient supply (i.e., each TN), there is a pair of numbers that represent the levels to which this species, growing in a monospecific stand, can reduce the concentration of the limiting soil nutrient and light at the soil surface. By graphing this pair of numbers for each rate of nutrient supply (over a wide range of nutrient supply rates) it is possible to construct a resource-dependent zero net growth isocline for this species (Fig. 3.7). Indeed, there is a family of isoclines that could be obtained, with each isocline showing the light level at a different distance above the soil surface. All such isoclines could be used to construct a 3-dimensional surface that would show the levels to which this species could reduce the limiting soil nutrient and light at all heights above the soil surface.

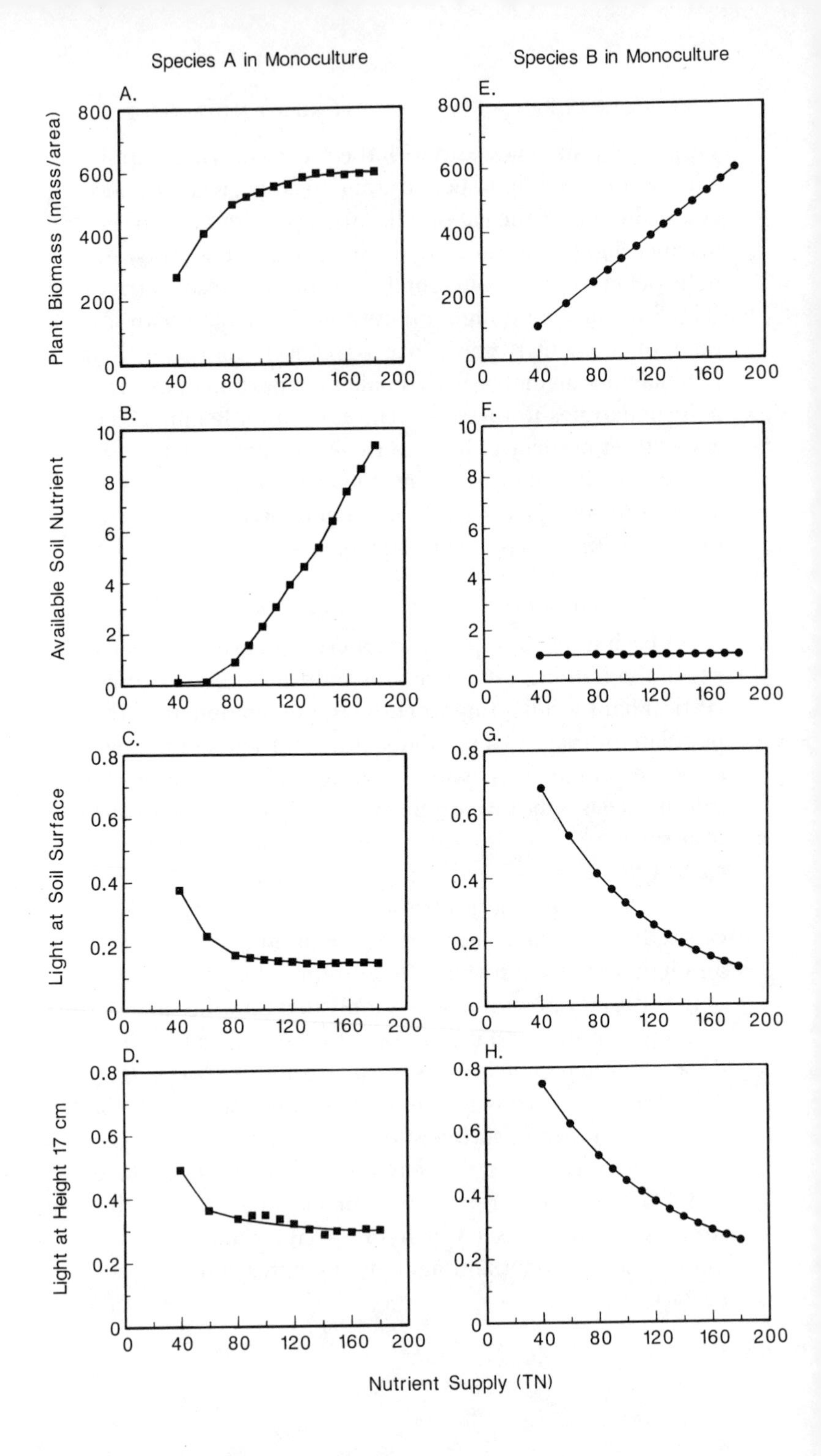
Species A in Monoculture
Species B in Monoculture
A.
E.
B.
F.
C.
G.
D.
H.
Plant Biomass (mass/area)
Available Soil Nutrient
Light at Soil Surface
Light at Height 17 cm
Nutrient Supply (TN)
0
40
80
120
160
200
800
600
400
200
10
8
6
4
2
0.8
0.6
0.4
0.2

Let us first consider the isocline that is based on the assumption that light intensity at the soil surface is more important than light intensity at any other height. The isocline of Figure 3.7 shows the levels of light and soil nutrient required for the long-term reproductive rate of this species to balance exactly its long-term mortality rate. If nutrient and light levels fall outside the isocline, the population size of this species should increase. If they fall inside the isocline, population size should decrease. The irregularities in the isocline are caused by the tendency for nutrient and light levels to oscillate. The isocline, remember, is based on long-term average resource availabilities.

Curved Isoclines

This isocline does not have a perfect right-angle corner, but rather has a curved corner. This means that the limiting soil nutrient and light are functioning as interactively essential resources. This is occurring despite the assumption that only the nutrient or light will control the rate of photosynthesis and that this species has a fixed pattern of allocation to roots, stems, and leaves (and thus cannot adjust its morphology to be equally limited by both resources). How can this occur?

This resource-dependent growth isocline is a summary curve for an entire size-structured population. The individual plants differ in their height and are living in a habitat with a vertical light gradient. Thus, some individuals are shorter and receive less light while other individuals are taller and receive more light. All individuals are assumed to

---

FIGURE 3.6. (A–D) When growing in monospecific stands at a loss rate of 0.2 wk^{-1} (in the computer model ALLOCATE), the long-term average ("equilibrial") biomass of species *A* and the average level to which it reduces the limiting soil nutrient and light penetration to the soil surface or to 17 cm above the soil surface depend on the nutrient supply rate. (E–H) Species *B* responds similarly to the nutrient gradient when growing in monospecific stands. The traits of species *A* and species *B* are given in the following section titled "Interspecific Competition."

experience the same, homogeneous soil. At intermediate resource supply rates, the shorter individuals are light limited and the taller individuals are nutrient limited. Thus, the shorter individuals within the population could be light limited at the same time and in the same habitat for which taller individuals are nutrient limited. The total effect of this dual limitation is a curved isocline. Two nutritionally perfectly essential resources could also function as interactively essential resources in a spatially heterogeneous habitat if individual plants are limited by different resources depending on their spatial location. Thus, vertical structure in light availability, horizontal spatial heterogeneity in a limiting soil nutrient, or morphological or physiological plasticity in acquiring resources could each cause nutritionally perfectly essential resources to function, in a habitat, as interactively essential resources. Data collected in a two-

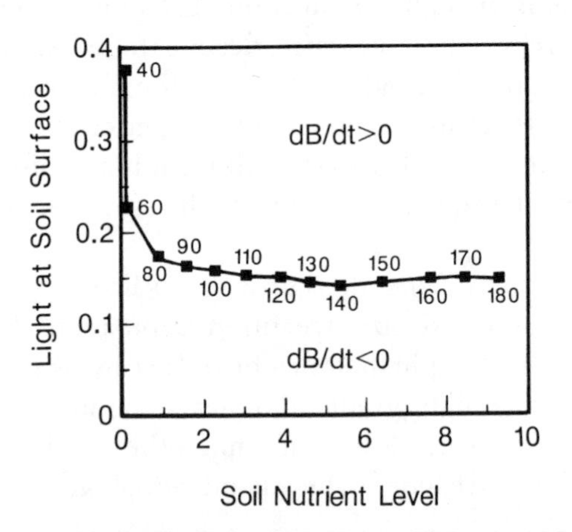

FIGURE 3.7. Resource-dependent growth isocline for species *A*. When ALLOCATE was solved with species *A* growing by itself to equilibrium, each different nutrient supply rate (TN) gave a different long-term average level of available nutrient and light penetration to the soil surface. The average resource levels associated with each TN are shown by a square, labeled with the TN value. These define a zero net growth isocline for this species in habitats with a loss rate of 0.2 wk^{-1}. The traits of species *A* are given in the following section on "Interspecific Competition."

year growth experiment at Cedar Creek Natural History Area showed that *Schizachyrium scoparium* (little bluestem, a native prairie grass) had a curved isocline (Fig. 3.8). The data for its isocline are based on the seasonal average level to which it reduced extractable NH_4 and NO_3, and the degree to which light penetrated through its leaves to the soil surface. Other data collected on its morphological plasticity showed that its root:shoot ratio ranged from 1.6 on the most nitrogen-poor soils to 0.7 on the most nitrogen-rich soils (Fig. 3.8B). Thus, its curved isocline is consistent with its ability to adjust its morphology in response to nitrogen and light availabilities.

Interspecific Competition

There are an almost unlimited number of distinct cases of competition among two or more species that could be explored using this model. I cannot discuss them all. Let us first consider several cases of competition among two species. What might happen if species *A*, which has been discussed above, competed with a second species, species *B*, that differed only in its half-saturation constants for nutrient- and light-limited growth? Thus, species *A* and *B* have the same pattern of allocation to leaves (AL = allocation to leaves = 0.7), to stems (AS = allocation to stems = 0.1) and to roots (AR = allocation to roots = 0.2). Both species have maximal heights of 200 cm, a seed size of 0.001 g, a maximal photosynthetic rate of 2.0 wk^{-1}, and respiration rates of roots, stems, and leaves of 0.3 wk^{-1}. Both species thus have identical RGR_{max}. The two species differ in their half-saturation rates for nutrient- and light-dependent photosynthesis (nutrient half-saturation constant for species *A* of KN(*A*) = 0.05; KN(*B*) = 0.50; light half-saturation constant of KL(*A*) = 0.5; KL(*B*) = 0.05) and in their leaf, stem, root, and seed nutrient concentrations (these are 0.01 for all four characteristics for species *A* and 0.02 for all four characteristics for species *B*). The parameters chosen give

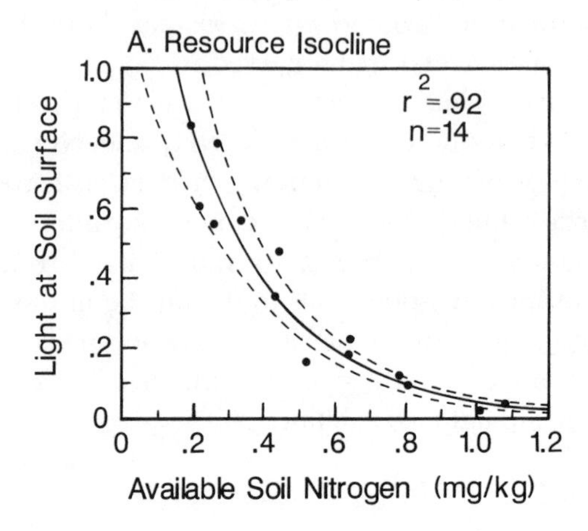

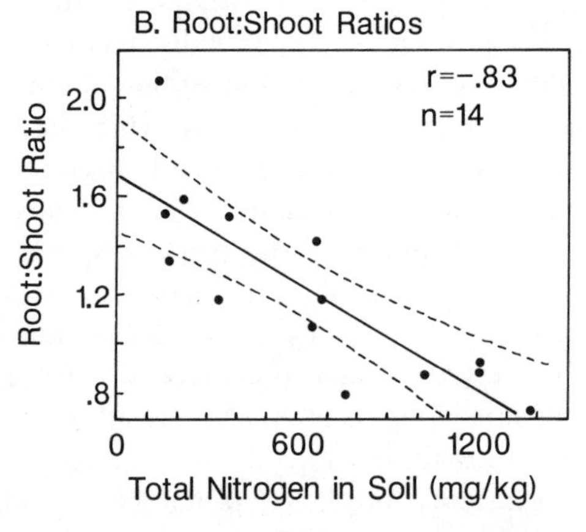

Figure 3.8 (A) Experimentally obtained zero net growth isocline for *Schizachyrium scoparium*. Each point is the average level of available nitrogen and of light penetration to the soil surface in an 18-liter pot of *Schizachyrium* grown under field conditions for two years. The 14 pots differed in the nitrogen supply rate of their soils. An exponential model was fitted to the data, giving $L = \exp(0.55 - 3.68N)$, with $r^2 = 0.92$, $P < 0.001$, where L is proportional light penetration to the soil surface and N (nitrogen) is the sum of KCl extractable ammonia and available nitrate, expressed as mg of nitrogen extracted per kg of dry soil. The dashed lines show the 95% confidence intervals for the regression. (B) Root:shoot ratios

species *A* a lower requirement for nutrient but a higher requirement for light than species *B* (Fig. 3.9A).

The eventual outcome of competition among these two species is determined by TN, the nutrient supply rate of the habitat. At low rates of nutrient supply (TN < 80), species *A* competitively displaced species *B*, as shown by Figure 3.9B. At intermediate rates of nutrient supply (80 < TN < 160), both species persisted, stably coexisting, although not at fixed population densities. At high rates of nutrient supply (TN > 160), species *A* was displaced by species *B* (Fig. 3.9B).

This pattern is qualitatively consistent with expectations based on the zero net growth isoclines of the two species (Fig. 3.9A). At low rates of nutrient supply, each species, when growing by itself in monospecific stands, reduces nutrient levels down to a point on the ascending (parallel to the y-axis) branch of its isocline, i.e., to a point for which each species is mainly limited by the soil nutrient. When these two species compete on a poor soil, species *A* should competitively displace species *B* because species *A* can reduce soil nutrient levels below those required for the long-term survival of species *B*. To see this, compare the resource trajectories of each species for TN = 60 in Figure 3.10. For TN = 60, species *A* reduces the amount of available nutrient to about 1/10th the level required for survival of species *B* at equilibrium. For very high rates of nutrient supply, both species are limited by light, and species *B*, which has the lower requirement for light at the soil surface, competitively displaces species *A* (see resource trajectories for species *A* and *B* for TN = 180 in Fig. 3.10). Coexistence occurs at intermediate rates of nutrient supply, for which species *A* is relatively more light limited and species *B* is relatively more nutrient limited. When the two species

---

for plants in these 14 pots when they were harvested at the end of their second growing season. A linear regression and its 95% confidence intervals are shown. Unpublished data of M. L. Cowan and D. Tilman, based on data collected in 1986.

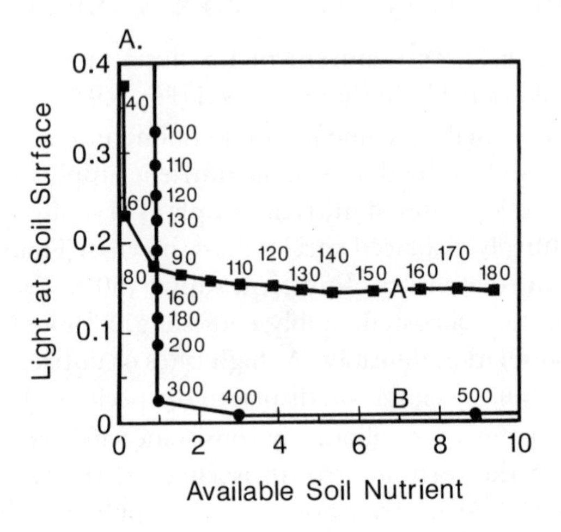

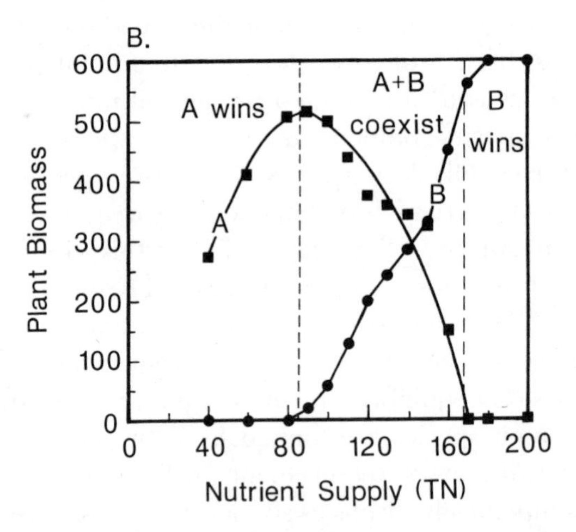

FIGURE 3.9. (A) Isoclines for species *A* and for species *B*, each one labeled to show the average levels to which each reduces resources for each nutrient supply rate (TN). Nutrient supply rates are shown next to the appropriate dots. Isoclines were derived using ALLOCATE, with each species growing in monospecific stands at different nutrient supply rates. (B) Long-term average results of interspecific competition between species *A* and *B*. *A* competitively displaced *B* for TN < 80, both species persisted for 80 < TN < 160, and *B* displaced *A* for TN > 160. This agrees with predictions based on the isoclines in part A. Biomasses shown are based on averages for the final 5 years (yr 25 to 30) of simulations of competition.

coexist, they reduce resource levels down, on average, to the point at which their isoclines cross, i.e., down to the two-species equilibrium point. Thus, even though ALLOCATE includes much more of the details of the mechanisms of plant competition for nutrients and light, and even though interactions never go to equilibrium, its qualitative predictions are similar to those that would be obtained from the graphical, isocline approach presented in Chapter 2. This qualitative similarity means that information on the long-term average resource requirements of these species (i.e., the shapes and positions of their zero net growth isoclines) obtained from single species monocultures can, in theory, predict the qualitative outcome of interspecific competition. If this were a general result, it would mean that it would not be necessary to obtain all the physiological and morphological details contained in the model ALLOCATE, but, instead, would be possible to summarize the morphological and physiological complexity of plants in an isocline that could be experimentally obtained from a series of single species monoculture gardens. However, further theoretical and experimental work is required to determine how useful an approximation the isocline approach may be.

Is there, though, a quantitative agreement between predictions based on single-species monocultures and the outcome of interspecific competition obtained from the model ALLOCATE? The average levels to which each species reduces nutrients and light at the soil surface when growing in monoculture can be used to predict the expected outcome of interspecific competition for any nutrient supply rate (TN). The approach that follows is the best approach of which I am aware for using monoculture equilibrial nutrient and light data to predict the outcome of competition. This is so because the directional aspect of light means that its supply rate and its consumption rate do not fit the mathematical definitions of Chapter 2. The graphical, equilibrium theory of resource competition summarized in Chapter 2 shows

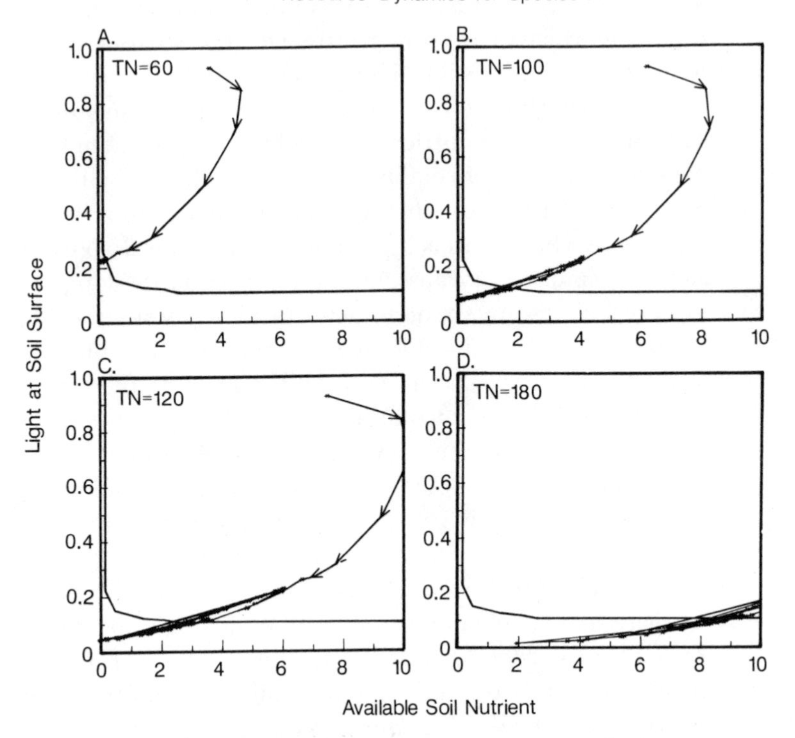

FIGURE 3.10. (A–D) The dynamics of resource consumption by species A, as predicted using ALLOCATE for monospecific stands at four different nutrient supply rates (TN). The unmarked curve is the net zero growth isocline for species A. The arrows show resource dynamics through time. (E–H) The dynamics of resource consumption by species B, shown as for species A, above. If the graphs for the two species for TN = 120 were superimposed, the reason for the

how isoclines and equilibrial resource levels obtained in monocultures can be used to predict the outcome of competition. This is illustrated graphically in Figure 3.11. Consider three different supply points, labeled x, y and z. Supply point x represents a habitat with a low supply rate of R_1 and a high supply rate of R_2. Supply point y represents a habitat with intermediate rates of supply of both resources, and supply point z represents a habitat with a high rate of

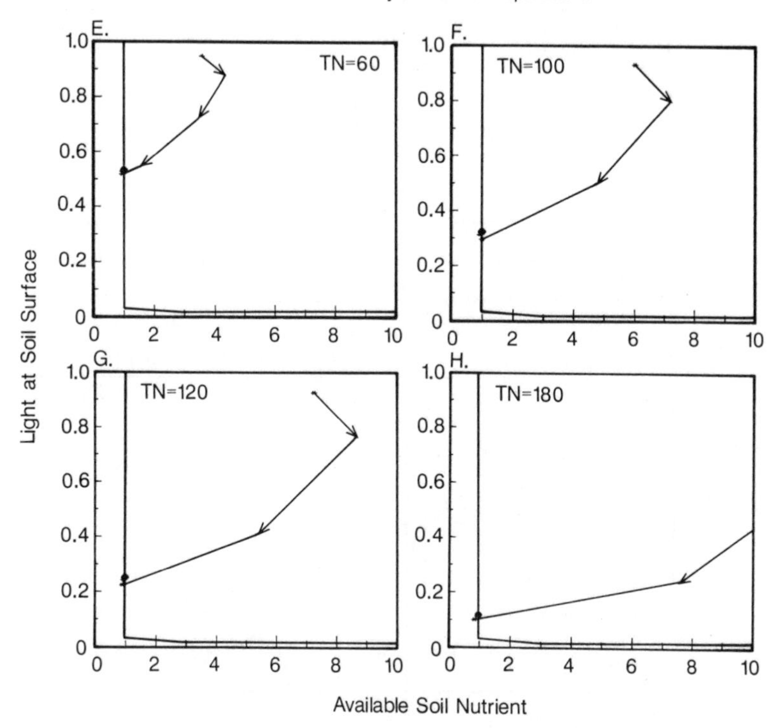

stable coexistence would be easily seen. Species A drives resource availabilities toward a point on its isocline for which there are sufficient resources remaining for species B to survive. Species B drives resource levels to a point on its isocline for which there are sufficient resources remaining for A to survive. Thus, for $80 < \mathrm{TN} < 160$, each species tends to inhibit itself more than it inhibits the other species.

supply of R_1 and a low rate for R_2. For supply point x (Fig. 3.11A), a monospecific stand of species L would reduce resource levels at equilibrium down to point $R^*_{L,x}$ on its isocline. Similarly, species M, when growing in a monospecific stand, would reduce resource levels down to point $R^*_{M,x}$. For supply point x, the equilibrial resource levels for each species are both on the same side of the two-species equilibrium point. Species L is predicted to competitively displace

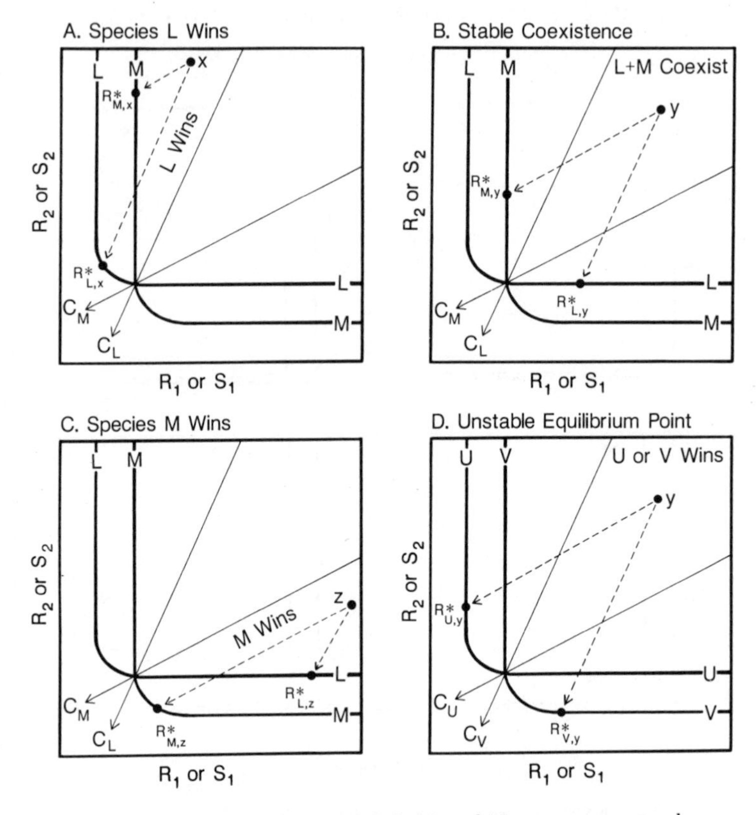

FIGURE 3.11. The thick curves labeled L and M are zero net growth isoclines for species L and M. (A) For supply point x, species L reduces resource levels, at equilibrium, to $R^*_{L,x}$. This is illustrated by the broken-line arrow. Species M reduces resource levels to $R^*_{M,x}$. In competition, species L displaces species M for all supply points in the region labeled "*L* Wins." (B) Species L and M can only coexist for supply points, such as point y, for which each species, at equilibrium, reduces resource levels down to a point that is in the region of net growth of the other species. Points $R^*_{M,y}$ for species M and $R^*_{L,y}$ for species L are such points, each associated with resource supply point y. (C) Species M competitively displaces species L for supply points in the region labeled "*M* Wins." Note that one such supply point, z, leads to R^*'s on the same side of the two-species equilibrium point. (D) The two-species equilibrium point is unstable if each species has an R^* that falls in the region in which the other species is unable to survive. Note that the only difference between this case and that of part B of this figure is the consumption vectors and the associated R^* values for the two species. Any supply point, such as y, that falls in the region labeled "*U* or *V* Wins" would lead to dominance by U or by V, with the other species competitively displaced. The winning species would be determined by initial conditions.

species M from such habitats because it can reduce R_1 to a level lower than that required for the existence of species M. Similarly, for supply points in the region in which species M is predicted to displace species L, such as point z (Fig. 3.11C), both $R^*_{L,z}$ and $R^*_{M,z}$ are on the same side of the two-species equilibrium point and species M is predicted to win because it can reduce R_2 to levels lower than species L requires to survive. It is only for resource supply points that fall in the predicted region of coexistence, such as for supply point y (Fig. 3.11B), that the equilibrium resource levels for each species in monoculture ($R^*_{L,y}$ and $R^*_{M,y}$) would fall on different sides of the equilibrium point. Note that this is a case of stable coexistence, and the monoculture equilibrium resource levels fall such that species L does not reduce resources below the levels required for the survival of species M, and that species M does not reduce levels below those required for the survival of species L for resource supply point y. This is because $R^*_{L,y}$ is in the region of net growth for species M and $R^*_{M,y}$ is in the region of net growth for species L.

The final case that can be considered is one in which there is an unstable equilibrium point (Fig. 3.11D). To distinguish this case from those of Figure 3.11 A-C, I have labeled the two species involved U and V. If an equilibrium point is unstable, each species, when growing in monoculture in an intermediate habitat, can reduce resource levels to points on opposite sides of the two-species equilibrium point, but these levels are such that each species leaves insufficient resources for the survival of the other. For example, supply point y in Figure 3.11D leads to $R^*_{U,y}$ and $R^*_{V,y}$. At $R^*_{U,y}$, only species U can survive, whereas at $R^*_{V,y}$, only species V can survive. If species U and V, when growing in monocultures, had isoclines and R^*'s as shown here, the predicted outcome of their competition would be unstable coexistence, i.e., either species U or species V completely displacing the other, dependent on starting conditions. In total, the analyses presented in Figure 3.11 suggest that empirically observed isoclines and the R^*'s associated with each species when growing in monoculture in habitats with particular resource supply characteristics should be sufficient information to predict the equilibrial outcome of interspecific competition. The underlying dynamics of resource supply and consumption then become unimportant except in how they influence isocline shape and position and R^*'s. This approach thus is a good way to analyze competition for light because it overcomes the difficulties of defining a "supply rate" for light and includes, in it, the inherent dependence of light consumption on total plant biomass.

These quantitative predictions of the graphical, equilibrium theory of resource competition can be applied to the cases of competition performed using ALLOCATE at various values of TN for species A and B of Figure 3.9. The isoclines of the two species, and the values of available soil nutrient and light at the soil surface to which each species reduced nutrients at several different nutrient supply rates (TN levels) are shown in Figure 3.9A. For TN from 40 to 80, both species A and B reduced resource levels, on average, to points that were above the two-species equilibrium point. Both species were mainly limited by the soil nutrient, and species A, which has the lower requirement for it, is predicted to win. The isoclines cross at the point on the isocline of species A for which it received, in monoculture, TN $\approx$ 82. Similarly, the isoclines cross at the point on the isocline of species B for which it received TN $\approx$ 150 (Fig. 3.9A). For all values of TN between 82 and 150, the nutrient and light levels of the single-species monocultures were on opposite sides of the two-species equilibrium point, and were such that coexistence was predicted to be stable. To see this, compare Figure 3.9A with Figure 3.11. This prediction, though, is based on the assumption that *light at the soil surface is a good measure of the total light gradient.* The simulations of competition for TN = 90, 100, 110, 120, 130, 140, and 150 all showed stable persistence of species A and B. The stability of the persistence was judged by doing numerous simulations for each TN, with each simulation starting at a different initial abundance of the two species. The isoclines predict that species *B* should displace *A* from all habitats with TN > 150. In such habitats, both species would be mainly light limited, and species *B* would reduce light at the soil surface down to a point below that required for the survival of species *A*. This did not occur at TN = 160. At TN = 160, *A* and *B* stably persisted. However, for TN > 160, *B* displaced *A* as predicted. Thus, these simulations have shown a close, but by no means perfect, correspondence between the predictions of the simple, equilibrium-based isocline approach and the outcomes of competition predicted by a more realistic and more complex model of competition for a soil nutrient and light. Thus, it may be possible for isoclines based on the long-term average levels to which plants reduce soil nutrients and light at the soil surface in monospecific stands to predict much of the pattern of the long-term outcome of pairwise competition.

Although this may seem to justify the use of light at the soil surface as an index of the degree of light availability throughout the entire vertical light gradient, the case discussed above includes a critical assumption. Both species were assumed to have the same maximal height at maturity (200 cm). What would happen if they had different maximal heights? To determine this, I kept all the parameters used in the case above constant, except that I modified the maximal height at maturity of species *A* to be 50 cm, and thenceforth called it species *C*. The height at maturity of species *B* was 200 cm, it being identical in every way to species *B* of Figures 3.5 through 3.11. This change in its maximal height moved the isocline of species *A* to the position shown for species *C* in Figure 3.12 and modified its competitive interactions with species *B*. The qualitative pattern of competition was not affected. Species *C* was dominant on soils with low rates of nutrient supply, the two species stably coexisted on soils of intermediate nutrient richness, and species *B* was dominant on rich soils. However, the isocline based on light at the soil surface was not as good at predicting the actual boundary between stable coexistence and dominance by species *B* in this case as in the case in Figure 3.9. The predicted region of coexistence based on light at the soil surface was from TN = 90 to TN = 165 (see Fig. 3.12A). However, the actual region of coexistence observed in the simulations was from TN = 90 to TN = 130 (Fig. 3.12B). Species *B* displaced species *C* for TN = 140 and all higher values tried.

This divergence from predictions based on the light at soil surface isocline is caused by the size structures of the populations. Species *B* is taller at maturity than species *C*. Its long-term average biomass increases with TN. For TN greater than about 130, it reduces light reaching the shorter, but mature, individuals of species *C* to a low enough point that these individuals become light limited and are not capable of producing enough seed to replace themselves. Thus, when plants have different heights at maturity, light levels other than light at the soil surface can be of critical importance in determining the outcome of competition. In this case, isoclines based on average light levels at some distance above the soil surface but below the tops of the shorter species (50 cm maximum height) are a much better predictor of the nutrient supply rates for which the species coexist. Isoclines based on light at 17 cm and 33 cm above the soil (Fig. 3.13) predict that the range of TN for which coexistence should occur is from about 90 to 130, very close to the region of coexistence observed in the simulations. However, I have not analyzed enough simulations to determine if this is a general result. I have, though, performed hundreds of simulations of competition among two or more species using the model ALLOCATE, and found that single-species monoculture-based isoclines were generally good predictors of the qualitative outcome of competition.

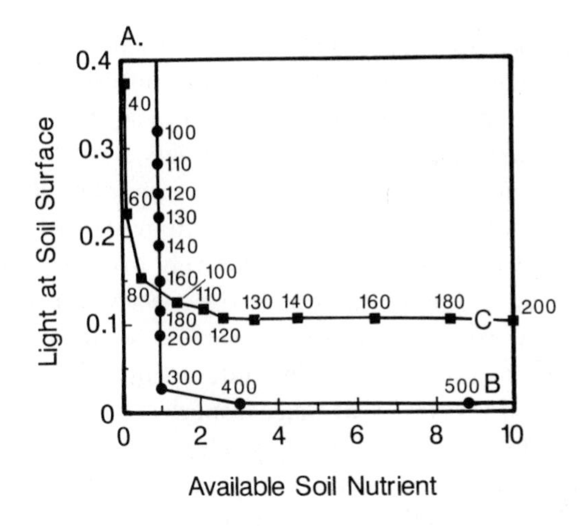

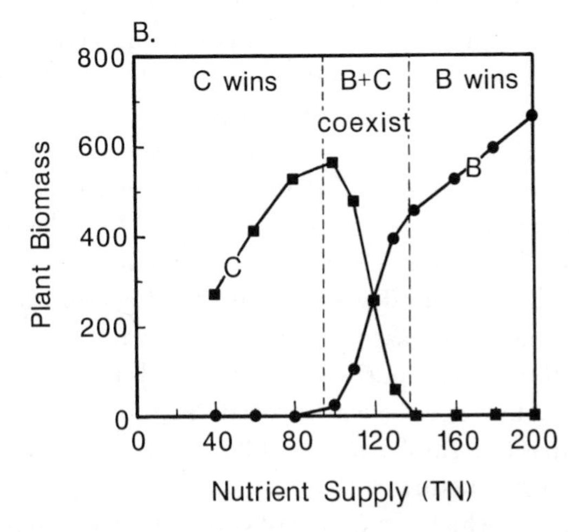

FIGURE 3.12. (A) Resource isoclines for species *B* (identical to species *B* of Figs. 3.6, 3.9, and 3.10) and for species *C* (identical to species *A* of those figures, except species *C* has a maximal height of 50 rather than 200 cm). (B) Outcome of competition between species *B* and *C*, as predicted by ALLOCATE.

The isoclines discussed above were based on the average levels to which each species reduced resources in monospecific stands. As shown in Figure 3.5, both light availability at the soil surface and nutrient availability actually had sustained oscillations in many of these cases. The long-term outcomes of competition discussed above, which the isocline models were able to qualitatively predict, were also based on average densities after the species had competed for 25 years. There are, however, some interesting dynamics to these competitive interactions (Fig. 3.14). At TN = 60, species *A* displaced species *B*, with a rather smooth, fairly asymptotic approach to the eventual "equilibrium." At TN = 200, *B* displaced *A* with a similarly smooth approach to equilibrium. In both of these cases, the winning species was nutrient limited. Species *B* did have a period of rapid increase early on in the interactions at TN = 60, and was more abundant than species A during the first year. The coexistence of the two species, such as at TN = 100, was not coexistence at fixed densities. Both species and both resources had sustained oscillations, with nutrient and light and the densities of species *A* and *B* oscillating out of phase. Resource levels oscillated so that first species *A* was favored then species *B* was favored, etc. However, each species was able to increase when rare, because high abundance of the other species pushed resource levels into a region in which the first species was favored. Such oscillations have not occurred in similar models I have solved that lacked an explicit size structure.

ALLOCATE, which is a model of competition among a size-structured population living in a horizontally homogeneous patch, predicts that one common dynamic interaction between two coexisting plants will be for each of them to change resource availabilities in ways that favor the other, and thus for each to replace the other on any small patch of soil. Just this pattern of local site replacements has been reported for beech (*Fagus grandifolia*) and sugar maple (*Acer*

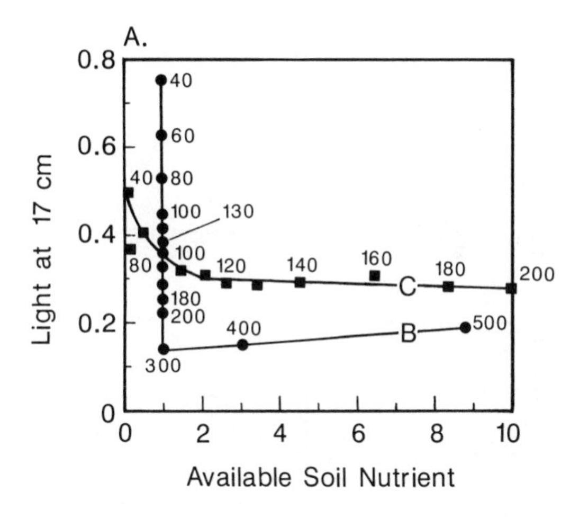

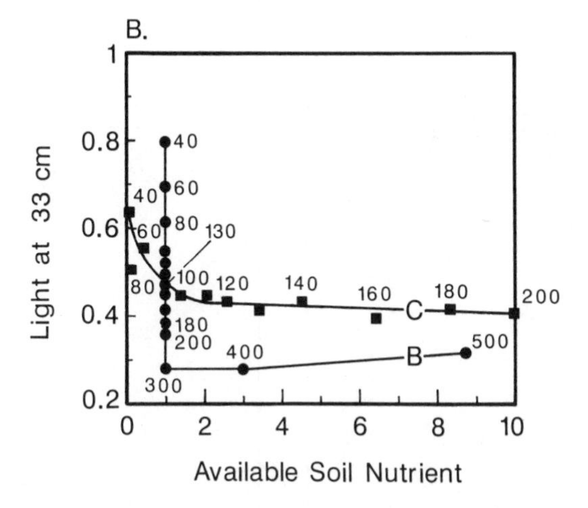

FIGURE 3.13. (A) Isoclines for species *B* and *C* based on light penetration to 17 cm. (B) Isoclines based on light penetration to 33 cm. In both cases, these isoclines are better at predicting the long-term outcome of competition between a tall species (species *B*) and a short species (species *C*) than isoclines based on light at the soil surface. To see this, compare this figure with Figure 3.12.

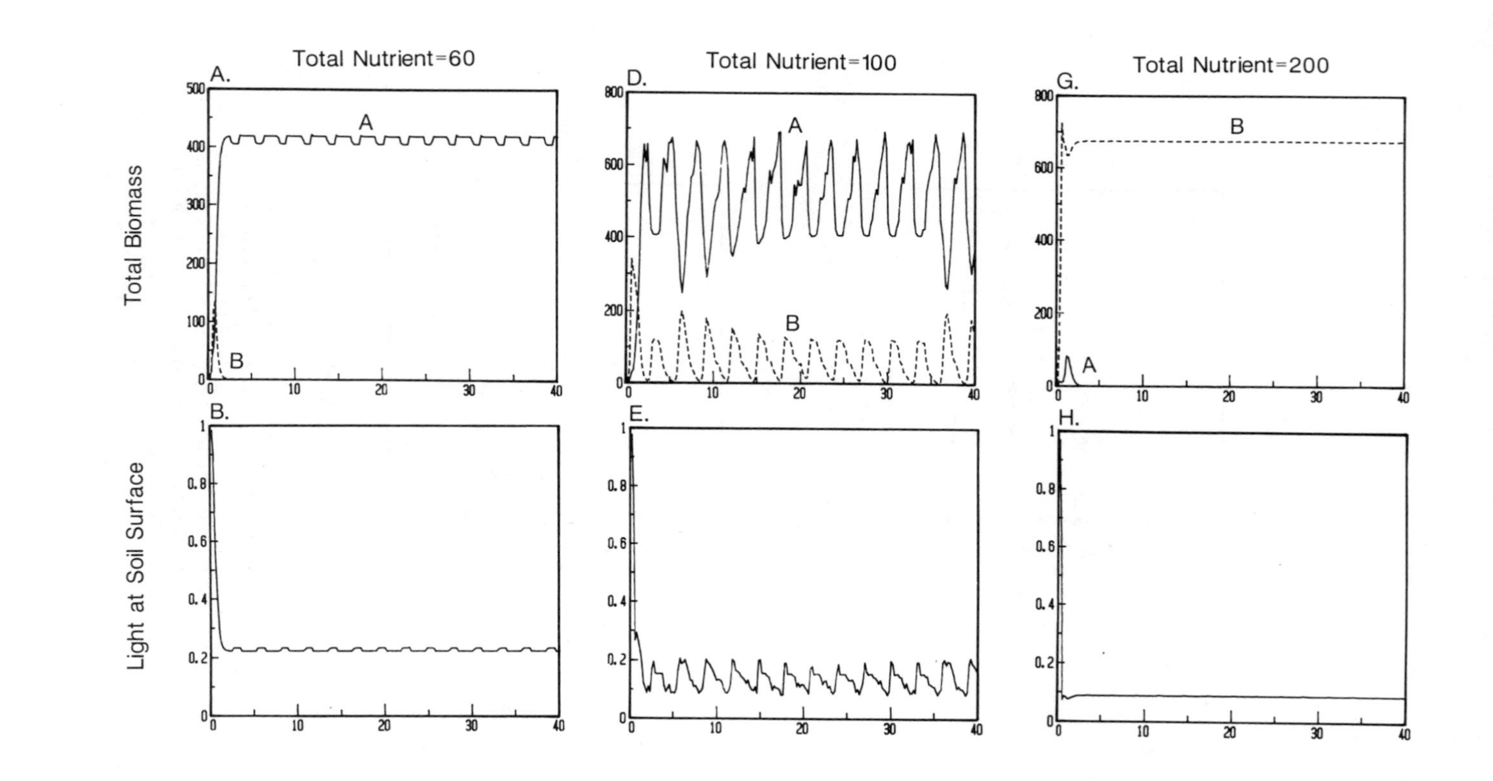
Total Nutrient=60
Total Nutrient=100
Total Nutrient=200
Total Biomass
Light at Soil Surface
A.
B.
D.
E.
G.
H.
A
B
0
100
200
300
400
500
600
800
0.2
0.4
0.6
0.8
1
10
20
30
40

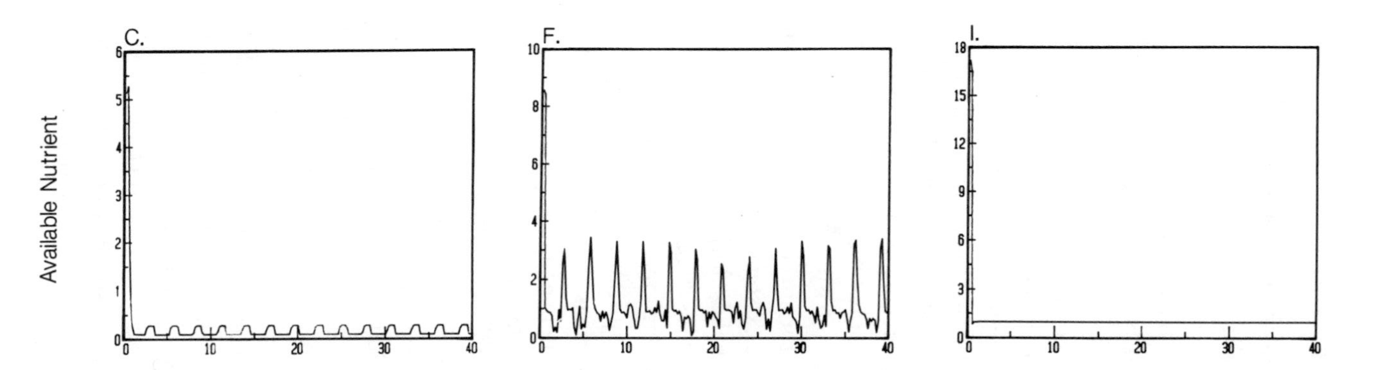

FIGURE 3.14. (A–C) Population and resource dynamics for competition between species *A* and *B* for TN = 60 as predicted by ALLOCATE. (D–F) Dynamics of competition between species *A* and *B* at TN = 100. (G–I) Dynamics of competition at TN = 200. Species *A* and *B* are identical to the species *A* and *B* of Figures 3.6, 3.9, and 3.10.

saccharum) in southwestern Michigan (Woods 1979; Woods and Whittaker 1981). Woods (1979) found that a maple tree is much more likely to have beech saplings under it than maple saplings, and that a beech tree is more likely to have maple saplings under it than beech saplings. Fox (1977) reported similar patterns for fraser fir and red spruce in the Great Smoky Mountains, between live oak and water oak in Florida, between beech and hemlock in Pennsylvania, and between subalpine fir and Englemann spruce in Medicine Bow Mountains (see Table 19.1 in Woods and Whittaker, 1981). All of these patterns are just what this model predicts should happen when two age-structured populations compete in an oscillatory mode in a habitat in which resource supply rates should lead to coexistence. It is within the realm of this model, though, for species to coexist without such oscillations. However, the existence of size structure in terrestrial vascular plants makes long-term oscillations quite likely on any given site. This suggests that the overall mechanism leading to coexistence may be the differentiation in the nutrient and light requirements of these species, and that the local dynamics, such as reciprocal replacement, are a result of the interaction between this mechanism and the size structure of these populations.

The theoretical examples of competition presented in this chapter have shown that the nutrient and light dependence of photosynthesis and the height at maturity of plants can influence competitive abilities. As will be shown in Chapter 4, allocation patterns, seed sizes, and size at first reproduction also influence competition among size-structured plants. Neither photosynthetic efficiency nor allocation to roots nor any other single physiological or morphological trait of plants can predict, by itself, what should happen when two or more size-structured plant species compete. The ability of a plant to grow and compete in a habitat with a particular resource supply rate and a particular loss rate is determined by interactions among numerous physiological, morphological, and life history traits.

These analyses suggested that experimentally obtained resource-dependent growth isoclines might serve as an acceptable summary of the total combined effects of the physiological, morphological and life history complexity of a species. Isocline shape and position are influenced by all of these plant traits. Further, the results suggested that the resource-dependent growth isoclines of individual species may be used to qualitatively predict the outcome of interspecific competition among terrestrial vascular plants. The isocline approach is an approximation because light is a complex variable. Some species may be more limited by light at the soil surface whereas other species may be more limited by light at some other vertical position. Individual members of a plant population may be limited by one resource when small and by another when large. Nevertheless, considering the differences in the complexities of the models and the number of parameters that must be estimated to use the isocline approach versus a model like ALLOCATE, the isocline approach may be an acceptable approximation for many field situations and for general theory. The ultimate usefulness of the resource-dependent isocline approach, or any theory, can only be determined by direct experimentation.

All models are abstractions—simplifications of the complexity of nature. The simplifying assumptions contained within a given model define the bounds within which it must be interpreted. Models that are not morphologically explicit can make no predictions about causes of morphological patterns. They should be used with caution in exploring the nature of the interactions among species that differ greatly in their morphology. The model of competition for nutrients and light among size-structured plants presented in this chapter will be used in the following chapter to determine how the morphologies and life histories of the plants that come to dominate a habitat should be influenced by the resource supply rate and the loss or disturbance rate of that habitat. The model ALLOCATE is one of many possible models of competition among size-structured plants. There are other constraints, tradeoffs, and simplifying assumptions that could have been built into it and there are insights to be gained by comparing such alternative formulations with ALLOCATE. I have discussed some of ALLOCATE's characteristics in this chapter to allow readers to become familiar with it and the effects of some of its assumptions. The model was not written to be a completely realistic model of the mechanisms of plant competition, but rather to include some potentially important aspects of competition among size-structured plants that are capable of continuous vegetative growth. Although there are many alternative ways that such models could be formulated, I believe that the major predictions made by ALLOCATE are robust and will not be contradicted by other size-structured models that also explicitly include the constraint of growth being determined by the pattern of allocation to roots, stems, and leaves.

SUMMARY

The maximal rate of vegetative growth of a plant, measured as $dB/dt \cdot 1/B$ in a habitat with unlimited resources, and often called its maximal relative growth rate, should be highly dependent on the morphology of the plant. A plant with a given nutrient- and light-saturated rate of photosynthesis and a given respiration rate can achieve its greatest rate of vegetative growth when it maximizes its allocation to leaves. Assuming that leaves, roots, and stems have approximately equal respiration and turnover rates, the maximal growth rate of a plant should be directly proportional to its allocation to leaves. The evolution of roots and stems must thus be explained as an adaptation to resource limitation.

ALLOCATE, an explicit model of growth and competition among size-structured plants, was presented and some of its features explored. Although ALLOCATE includes such complexities as (1) nutrient- and light-dependent photosynthesis, (2) continuous, resource-dependent, allocation-determined vegetative growth of plants, (3) distinct root, leaf, and stem biomass and respiration, and (4) a vertical light gradient that is imposed by the plants themselves, many of the long-term outcomes of the model can be qualitatively predicted using nutrient- and light-dependent growth isoclines. Such isoclines are obtained from the model by allowing each species to grow by itself in monospecific stands. The size structure that is included in the model makes many populations, especially when light-limited, grow with sustained oscillations. Among its other predictions, the model suggests that two plants could coexist in a habitat with a constant loss rate and with a uniform soil by having seedlings of each species establish and grow better in the presence of adults of the other species. The coexistence occurs because one species is a better competitor for the soil resource and the other is a better competitor for light. The oscillatory nature of the interaction is caused by the size structure of the populations. The model presented in this chapter is used in Chapter 4 to explore the effects of plant morphology and life history on competitive interactions in differing habitats.

CHAPTER FOUR

Allocation Patterns and Life Histories

> It is likely that through a quantitative understanding of how different plants gain and allocate their resources it will be possible to make predictions as to their success in any given physical environment in combination with any competitor and predator.
>
> —H. A. Mooney (1972, p. 315)

As discussed in Chapter 3, any allocation to roots or stems comes at a cost—a decreased maximal rate of vegetative growth. Despite this cost, a wonderful array of plant morphologies has evolved. In this chapter I use the model of plant competition for nutrients and light developed in Chapter 3 to explore the role of soil resource supply rates and loss rates in the evolution of plant morphologies and life histories. It has long been recognized that there are strong correlations between plant physiognomy and such physical characteristics of plant habitats such as soil type and climate (e.g., Raunkiaer 1934; Billings 1938; Mooney 1972). Can these correlations be explained by the leaf-stem-root tradeoff that plants face? Does this tradeoff necessarily constrain each plant morphology to being a superior competitor for only a small region of the nutrient-supply-rate and loss-rate plane?

In this chapter, I explore numerous cases of competition using the model ALLOCATE. Although I will compare some of these to the isocline approach, the main purpose of the following sections is to determine broad, general relations that may exist between plant morphology and plant competitive ability along soil-resource:light gradients and loss gradients. Other aspects of ALLOCATE, including the effects of allocation patterns on the dynamics of competition, are discussed later in this book. This chapter considers cases of multispecies competition for which the plant species (or phenotypes) are identical in all their traits except their allocation of photosynthate to leaves, stems, and roots. This is an important case, for it represents a tradeoff that all plants face.

ALLOCATION AND MORPHOLOGY

A plant's allocation pattern has a large influence on its morphology. The relative mass of leaves, roots, and stem for a plant is determined by the relative rates at which these are produced and lost. In this chapter, I assume that the only losses of plant parts are caused by respiration, and that leaves, roots, and stems have equal respiratory rates. Given these restrictions, allocation and morphology become synonymous. Clearly, though, they are not synonymous for most real plants. For most plants, the results presented in this chapter would be better interpreted as coming from differences in morphology, realizing that morphology is strongly influenced by allocation.

The plant allocation patterns or morphologies discussed in this chapter can be viewed in two different ways. First, each morphology (i.e., each unique allocation pattern) could be considered to represent a different species. Even though individuals within a species differ in their morphologies, the range of variation within a species is small compared to the range among species. Thus, as a first approximation, each species could be considered to have a particular, morphologically fixed life form, with the proportion of roots, leaves, and stems remaining fixed during vegetative growth from seedling to adult. Second, each morphology could be viewed as a different genotype within a given species, and the results presented in this chapter could be thought of as elucidating the habitats (as defined by resource supply rates and loss rates) in which each genotype would be favored. Both of these are suitable views of the mathematics underlying the results presented in this chapter. This suggests that the same qualitative pattern of morphological variation in relation to resource supply rates and loss rates is expected among genotypes of a particular species as is expected among different species. Moreover, it would seem logical to expect that the pattern of morphological plasticity within each individual plant would show a similar dependence on resource supply rates and loss rates. This latter possibility, though, is not discussed until Chapter 9.

ALLOCATION PATTERNS AND ISOCLINE POSITIONS

Let's start by considering six plants that are identical in all ways except their allocation patterns. Figure 4.1 gives their allocation patterns and shows their resource-dependent growth isoclines as derived from ALLOCATE. Allocation pattern clearly influences isocline shape and position. In general, individuals with a higher allocation to roots can survive at lower soil nutrient levels than individuals that allocate more to leaves and stems. To see this, compare the position of the isocline of species *A*, which has the highest allocation to roots, with the positions of the isoclines of the other species (Fig. 4.2A). Allocation to both leaves and stems influences the ability of an individual to survive in habitats with low light intensity at the soil surface.

FIGURE 4.1. Resource-dependent growth isoclines (predicted by ALLOCATE) for 6 species that are identical in every way except in their pattern of allocation to leaves (AL), to roots (AR) or to stems (AS). The allocation patterns are listed, as is the nutrient supply rate (TN) that led to various points on each of these isoclines. Note that the isoclines use light at the soil surface. Isocline position is influenced by root, leaf, and stem allocation patterns. Increased root allocation shifts the isoclines to the left, but upward, making species be better competitors for the soil nutrient but poorer competitors for light. Increased allocation to leaves and stems shifts the isocline

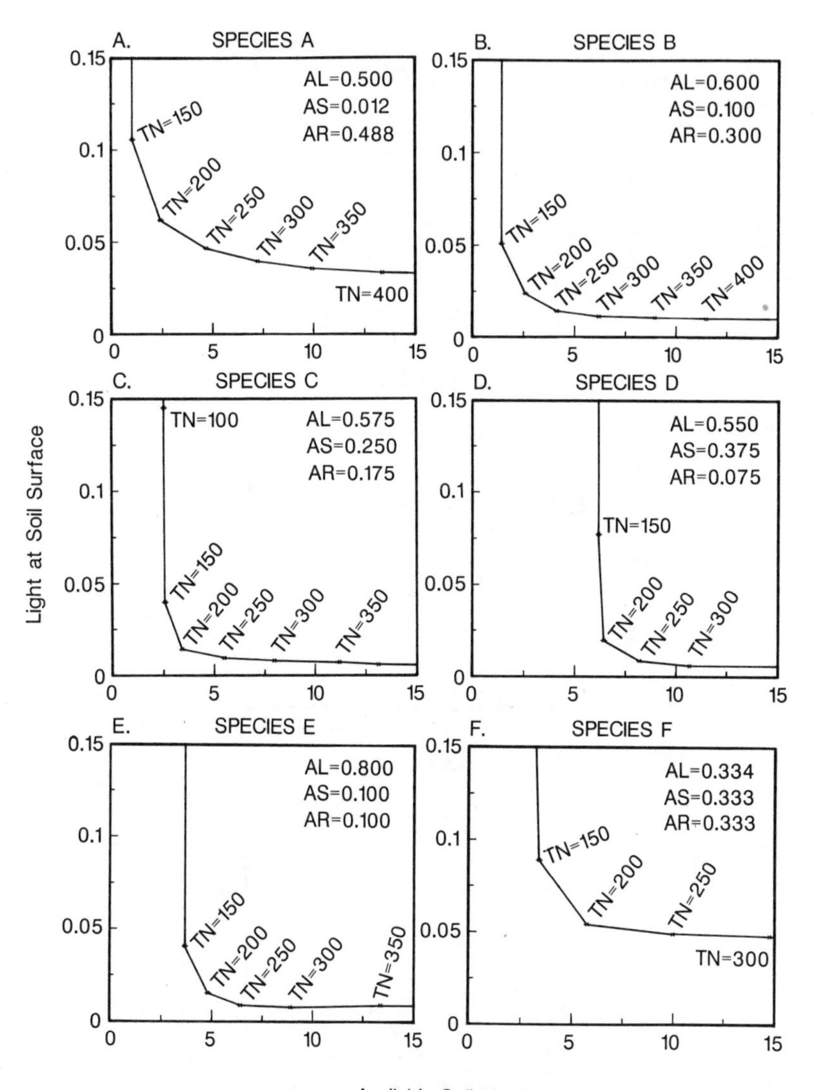

in the opposite direction, generally increasing competitive ability for light, but decreasing nutrient competitive ability. The six allocation patterns were chosen to give four species (species *A–D*) with allocation patterns that were competitively superior and two species (species *E* and *F*) that were not. These isoclines lead to two-species equilibrium points that are unstable equilibria.

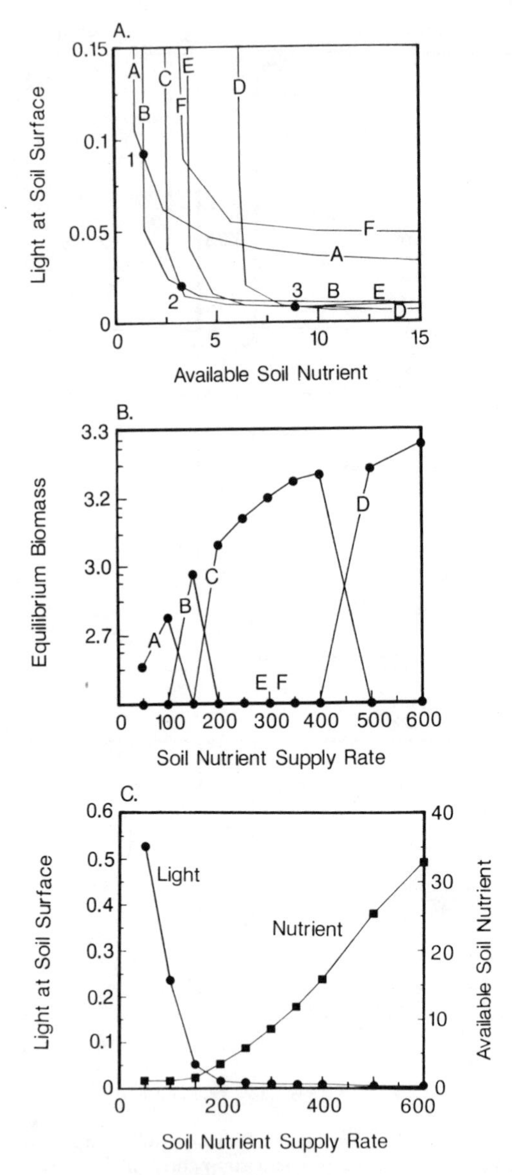

FIGURE 4.2. (A) The isoclines for species A–E of Figure 4.1, superimposed in the same figure. Dots show two-species equilibrium points, each of which is unstable. (B) The long-term outcome of competition among the 6 species of Figure 4.1, as predicted by

When individuals with these six different allocation patterns compete in a variety of habitats that are identical except in their rate of supply of the limiting soil resource, species *A* comes to dominate the most nutrient-poor habitats, species *B* dominates more nutrient-rich habitats, species *C* dominates even richer habitats, and species *D* dominates the habitats with the highest rates of nutrient supply (Fig. 4.2B). Species *E* and *F* are eventually displaced from all habitats, independent of the rate of supply of the limiting soil nutrient. All these results are based on the long term outcome of competition and are averages over any persistent oscillations. This model of competition among size-structured populations also shows that, in the long term, the average soil nutrient level increases and average light penetration to the soil surface decreases along such a productivity gradient (Fig. 4.2C).

The allocation differences among these six species do not lead to stable coexistence of different pairs of species along the productivity gradient (Fig. 4.2B). Rather, each of the two-species equilibrium points is locally unstable, as can be determined by the method of Figure 3.11D using the R^*'s shown for various TN levels in Figure 4.1. For instance, for TN = 150, the R^* for species *A* is above the equilibrium point labeled 1 in Figure 4.2A, and that for species *B* is below it. The R^* for species *A* indicates that it can reduce the light level at the soil surface and the soil nutrient level sufficiently for a habitat with TN = 150 that species *A* could displace all other species. However, the R^* for species *B* indicates that it could also displace all other species from a habitat with TN = 150. Thus, these species do not have a

---

ALLOCATE. Note that species *E* and *F* are displaced from all habitats along the nutrient supply gradient at this loss rate. Species *A*, *B*, *C*, and *D* become separated along this gradient in an order determined by their isoclines. There are no supply rates (TN) for which these species stably coexist. (C) Long-term average levels to which the soil resource and light at the soil surface are reduced during competition among these 6 species.

stable two-species equilibrium point, and the outcome of their interaction at TN = 150 should depend on initial conditions. In the simulations used to generate Figure 4.2B, both species started as seedlings with equal densities, and *B* displaced *A* at TN = 150. Species *B* and *C* also lack a stable two-species equilibrium point, as do species *C* and *D*. However, each species does have a range of soil resource supply rates (TN) for which it is the superior competitor independent of starting conditions. For instance, species *A* displaced all other species for TN less than about 100, species *B* displaced all other species for TN between about 160 and 180, species *C* displaced all others for TN between about 220 and 250, and species *D* displaced all other species for TN greater than about 500.

These results demonstrate that different patterns of allocation, in otherwise identical plants, do not lead to the stable coexistence of two species in a homogeneous habitat in which a soil resource and light are the limiting resources. This occurs because of the strong link between allocation patterns and resource consumption rates. When plants compete for two resources, two species can stably coexist only if each species consumes relatively more of the resource that more limits it at the two-species equilibrium point (Tilman 1982; see Fig. 2.8 or 3.11). However, higher allocation to one structure decreases the likelihood that the resource obtained by that structure will be limiting, because higher allocation increases the amount of that resource that is consumed relative to the other resource. For species that differ only in their allocation patterns, a species gains competitive ability for a resource by consuming a higher proportion of that resource. This means that, at a two-species equilibrium point, that species is unlikely to be limited by that resource, but it will consume a higher proportion of it. Thus, each species would tend to consume relatively more of the resource that did *not* limit it at the two-species equilibrium point, and the equilibrium point would be unstable.

Other factors, such as species that are superior nutrient competitors requiring lower tissue nutrient levels, or the tendency toward self-limitation because of the "neighborhood" aspect of competition, encourage stable coexistence of various pairs of species in a habitat in which a soil resource and light are limiting. Differential allocation of photosynthate to roots, leaves, and stems, however, does not.

These simulations demonstrate two important points. First, species with some allocation patterns, such as E and F, are not viable competitors at any point along a given soil-resource:light gradient. Second, the allocation pattern that is a superior competitor for one region of a soil-resource:light gradient is an inferior competitor at other points on the gradients. These results also suggest that the qualitative outcome of competition and the stability of equilibria can be predicted, in some cases, by the resource-dependent isoclines and the R^*'s of the species.

COMPETITIVELY SUPERIOR MORPHOLOGIES

To determine which allocation patterns will be competitively superior in particular habitats, it is necessary to allow competition among numerous individuals that differ in their allocation patterns (morphologies) but are otherwise identical. Each of these distinct morphologies, determined by its allocation pattern, could be considered to be a distinct species or could be considered to be a genotype within a parthenogenetic population. One simple alternative to performing such simulations might be to use optimization procedures, such as was done by Iwasa and Roughgarden (1984). Their work provided several major insights into the relations between morphology and habitat. However, such optimization approaches assume *fixed resource levels*, and thus do not include the process of resource competition. As Iwasa and Roughgarden (1984) noted, the optima depend on resource levels. In competitive situations, resource levels are determined by the competing species. Although the optimization approach can provide some insights, it does not include the feedback effect of growth on resource levels that is the force driving competition. Thus, I chose to use an explicit, dynamic model of resource competition to determine which morphologies would be superior competitors for particular environmental conditions.

For these simulations, all parameters except allocation were held constant. Thus, there were no stable, two-species equilibrium points. I first determined which allocation patterns were viable in the absence of interspecific competition, i.e., which allocation patterns could maintain viable populations (biomass > 0) in the absence of competition from species with different allocation patterns. To do this, I chose allocation parameters for 171 different species so as to have the species be uniformly spaced over the entire range of possible patterns of allocation to leaves, stems, and roots. Each of these different allocation patterns could be represented as a point in the allocation triangle of Figure 4.3. Note that a single point on this figure shows the proportional allocation to leaves (x-axis) and stems (y-axis). Because all new growth is allocated to either leaves, stems, or roots (until a plant reaches its height at maturity, at which point all photosynthate in excess of respiration is allocated to seeds), the proportional allocation to leaves, stems, and roots must sum to 1.0. This means that the proportional allocation to roots can be determined by summing the allocations to leaves and stems, and subtracting this total from 1. This is easily visualized graphically. Each of the diagonal lines in the allocation triangle of Figure 4.3 is a line along which root allocation is constant. Each point on the triangle of Figure 4.3 is a different plant morphology. Plants with low allocation to stem have a rosette life form, and range from rosettes with high root:leaf ratios (to the left) to rosettes with low root:leaf ratios (to the right). Those with

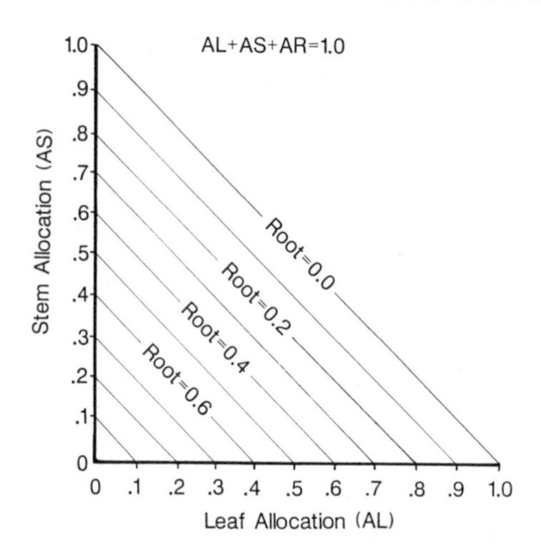

FIGURE 4.3. All possible patterns of allocation to leaves (AL), to stems (AS), and to roots (AR) can be shown on a graph like this. Every possible point within this "allocation triangle" represents a distinct plant morphology. Note that AL + AS + AR = 1, i.e., photosynthate is constrained to being allocated to leaves, stems, or roots. Each diagonal line is a line of equal allocation to roots. Plants with allocation patterns that fall near the origin have a high proportion of their biomass in roots. Those near the apex of the triangle have a high proportion of stem. Those near the right-hand corner of the triangle are mainly leaf.

high allocation to shoots (toward the apex of the triangle) have taller, more upright growth forms. Stem allocation determines how rapidly plants increase in height, and thus average plant height.

When each of these 171 species was allowed to grow by itself to equilibrium in a particular habitat, i.e., at a particular loss rate and nutrient supply rate, there were morphologies that led to inviable plants. These inviable plants could not maintain themselves in a habitat even in the absence of interspecific competition. These inviable morphologies are shown by the shaded regions in Figure 4.4. They could not survive because their loss rate, both from respiration and mortality, was greater than their growth rate. A wider range of allocation patterns was viable, in the absence of interspecific competition, on richer soils. The shape of the region of viable morphologies on poor soils (TN = 50) indicates that allocation to stems is relatively unimportant for survival in monoculture but that there is a minimum allocation to roots and leaves. This minimum level of allocation to leaves and to roots decreases as soil richness increases. On rich soils (TN = 1000) the boundary becomes almost a straight line with the proportional allocation to leaf being the most critical parameter. For the parameters used in these simulations, plants that allocated less than about 20% to leaves had insufficient leaf area for photosynthesis to exceed respiration and other losses. Clearly, the actual boundary between viable and inviable allocation patterns depends on respiration rates of each tissue type, mortality or loss rates, and the nutrient and light dependence of photosynthesis. These analyses indicate the qualitative relationships expected between allocation to roots, leaves, and stems, and the necessity for plants to have sufficient leaf mass to "pay the cost" of root and stem respiration. Moreover, these analyses suggest that there is a broad range of viable morphologies. How do these viable morphologies fare when they compete against each other?

Allocation and Competition on Productivity Gradients

To determine the outcome of competition among plants that differed only in their morphologies (allocation patterns), 200 different allocation patterns were chosen within the region of viable allocation patterns. All 200 morphs were then allowed to compete simultaneously with each other for a limiting nutrient and light, using the model ALLOCATE. Because of the size and complexity of the resulting model, it could only be solved using a Cray 2 supercomputer. All of the simulations ran for 40 "years." At any given rate of nutrient supply (TN) and any given loss rate, one or two morphologies were competitively domi-

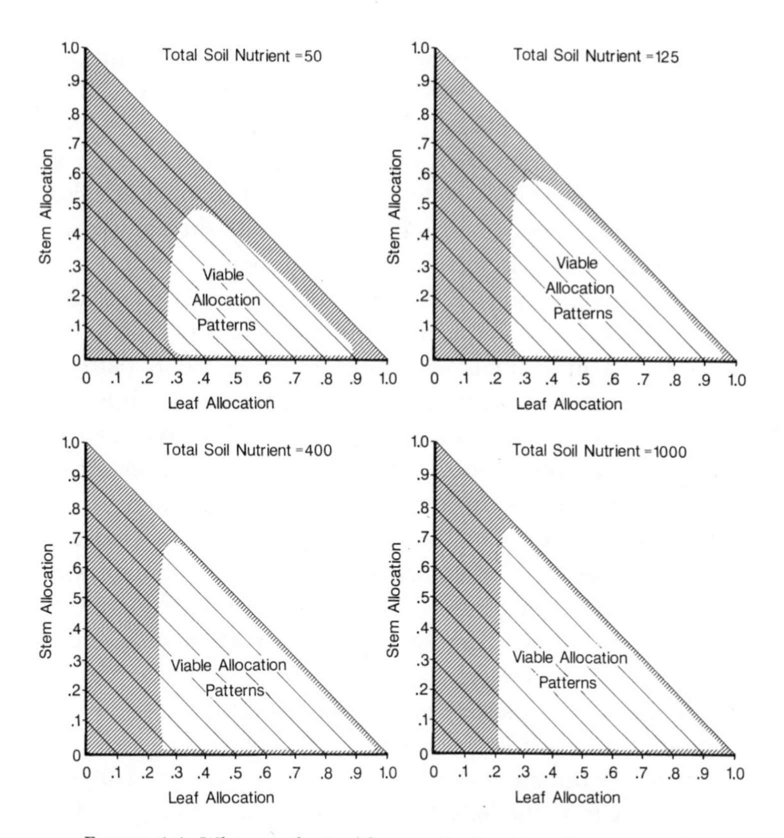

FIGURE 4.4. When a plant with a particular allocation pattern is allowed to grow in a monospecific stand, there are some allocations that lead to inviable plants, i.e., to plants that cannot maintain themselves in a particular habitat even in the absence of interspecific competition. All of the allocation patterns that lead to inviable populations fall in the shaded regions of this figure. An increasing range of morphologies are viable in habitats with higher rates of nutrient supply. For all cases, the loss rate was 0.2 wk^{-1}. Results are based on simulation using ALLOCATE. If photosynthetic rates were higher or if root and stem respiration rates were lower, the region of inviable morphologies would be smaller. These figures show which allocation patterns were viable for one particular set of parameter values—the parameters used for the simulations of Figures 4.5–4.8.

nant, displacing all the rest of the 200 species. If two species were dominant, they had very similar, usually adjacent, allocation patterns. This suggests that a single allocation pattern, or at least a small region of allocation patterns, leads to superior competitive ability in any given habitat. It also demonstrates, once more, that ecologically similar species can co-occur for long periods of time before competitive displacement occurs (cf. Hubbell and Foster 1986). I call the competitively dominant morphology, which is capable of displacing individuals with any other allocation pattern from a given habitat, the superior morphology or superior allocation pattern for that habitat. I found it necessary to use a large number of species (200) in order to obtain a "smooth" transition from one dominant morphology to the next along the productivity or loss gradients. This is at least partly a result of the unstable two-species equilibrium points, which can cause the competitively superior morphology to "jump" from one allocation pattern to another when there are few alternative allocation patterns.

By performing such simulations of multispecies competition at many different rates of nutrient supply, I determined how the superior allocation pattern depended on the productivity of the habitat. The curve traced in Figure 4.5 shows the morphologies that led to superior competitive ability at different points along this soil-nutrient to light gradient. In very unproductive habitats (TN = 25 and 50) the superior allocation pattern had high allocation to roots and leaves and low allocation to stems. In habitats with higher rates of nutrient supply (TN = 150), slightly greater allocation to stems was favored, but the major change was lower allocation to roots and more to leaves. Thus, in very nutrient-poor habitats, increases in nutrient supply rates favor plants that give up roots for leaves. Further increases in nutrient richness favor greater allocation to stem, mainly at the expense of allocation to roots. On very rich soils, further increases in productivity lead to increased allocation to stem mainly at the expense of leaves.

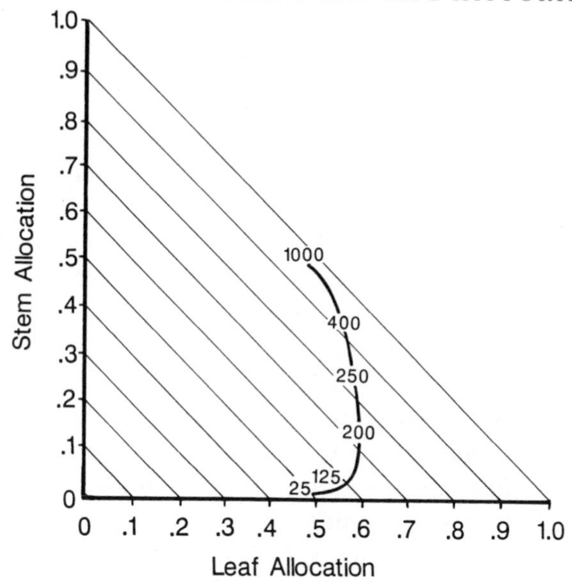

FIGURE 4.5. The competitively superior morphologies at a loss rate of 0.2 wk^{-1} are determined by the nutrient supply rate (TN), shown as numbers along the curve. This is based on competition among 200 morphs that were identical in all respects except their pattern of allocation. Different morphologies are competitively superior at different nutrient supply rates (TN). The TN values used range from 25 (an extremely nutrient-poor soil) to 1000 (a very rich soil).

These results demonstrate that the ability of a plant to compete for a soil resource and light is highly dependent on plant morphology, i.e., on the proportion of a plant's biomass that is allocated to leaves, stems, or roots. Of all the possible morphs, only one particular allocation pattern led to superior competitive ability in a particular habitat. Thus, a plant with a morphology that allows it to be a superior competitor at one nutrient supply rate is necessarily an inferior competitor at higher and lower supply rates compared to plants with other morphologies.

The response of plants to the different rates of supply of the limiting soil nutrient (Fig. 4.5) led to inverse gradients in available soil nutrients and light intensity at the soil surface (Fig. 4.6A). Total plant biomass and the soil resource level increased with the rate of nutrient supply whereas light penetration to the soil surface decreased. Allocation to roots decreased along this gradient, allocation to stems increased along the gradient, and allocation to leaves reached a maximum at intermediate rates of nutrient supply (Fig. 4.6B). The results summarized in Figures 4.5 and 4.6 were derived for one particular set of physiological constants, one set of loss rates, one height at maturity, etc., for all 200 morphs. I have performed two other sets of simulations like these (with different parameters, but with the species in each simulation being identical except for allocation patterns) at this same loss rate and found the same qualitative dependence of the superior allocation pattern on the rate of nutrient supply. The actual quantitative pattern of allocation depends on the costs of producing and maintaining each tissue type, on differences in height at maturity of plants dominant at different points along the gradient, on the structural constraint relating height to stem and leaf mass, and on differences in other physiological parameters, including the light and nutrient dependence of photosynthesis. However, although each of these can influence the quantitative pattern of allocation expected along a productivity gradient, the same general forces influence allocation patterns along all productivity gradients. Allocation to roots decreases with increased rates of nutrient supply. Allocation to non-photosynthetic, structural tissues is greater in more nutrient-rich but lower light habitats because rapid increases in height are of great importance in low light habitats. These two forces often cause allocation to leaves to be maximal in habitats of intermediate nutrient supply rates.

The different allocation patterns predicted for different points along the soil-resource:light gradient affect not only the competitive abilities of the species for the limiting resources, but also their maximal growth rates. The predicted maximal growth rate associated with each allocation

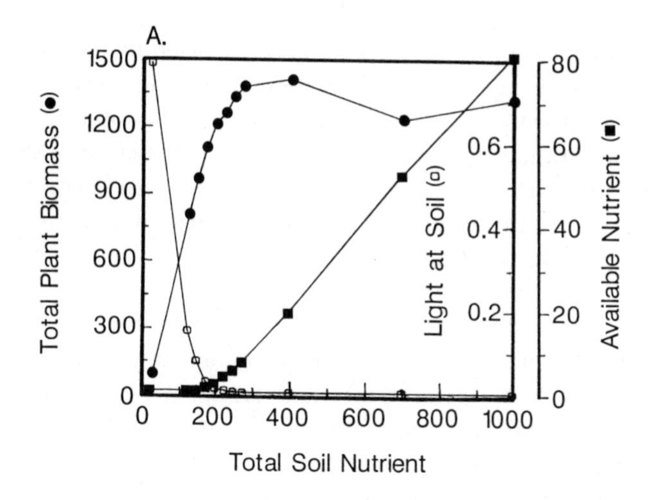

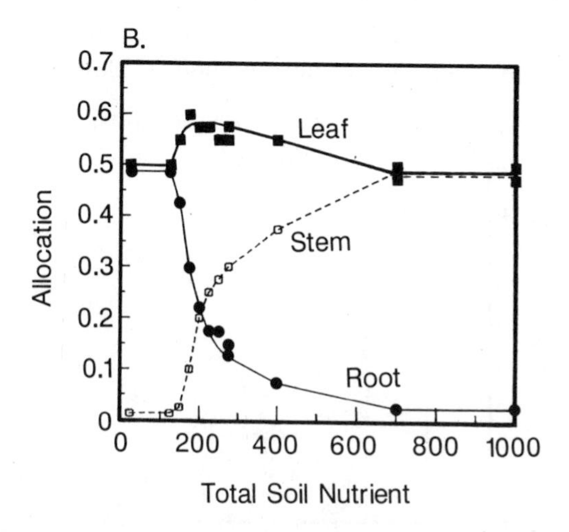

FIGURE 4.6. (A) Dependence of long-term average plant biomass (dots), soil nutrient level (squares), and light penetration to the soil surface (open circles) on nutrient supply rate for the case illustrated in Figure 4.5. (B) Leaf, root, and stem allocations of the competitively superior morphs as a function of the nutrient supply rate of a habitat, for the case illustrated in Figure 4.5.

pattern is shown in Figure 4.7. As predicted by the simple theory of Chapter 3 (Eq. 3.5), these maximal rates of vegetative growth closely correspond with the proportion of photosynthate that is allocated to the production of new photosynthetic tissues. Such differences in maximal growth rate have a profound effect on the population dynamics of these plants (see Chapter 6).

Morphologies on Loss Rate Gradients

Similar analyses were performed for habitats in which these same plants experienced different loss rates. The loss imposed was a density-independent mortality that fell equally on all individuals independent of their size or allocation pattern. For each loss rate, I determined which allocation pattern, of the 200 total patterns, was the superior competitor at each of numerous points along a soil-

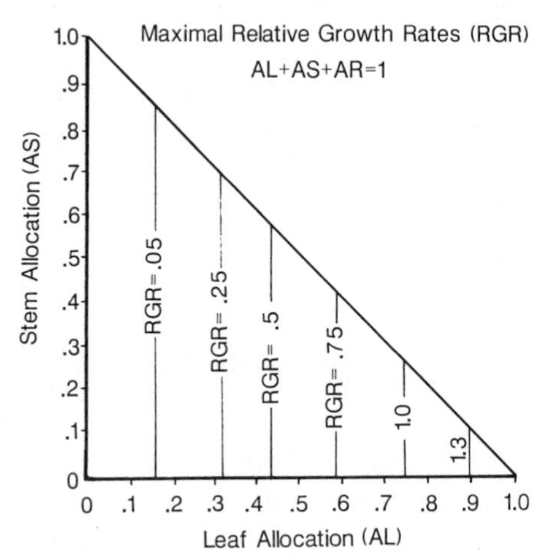

FIGURE 4.7. Maximal growth rates (RGR_{max}) of morphs that differ in their pattern of allocation to roots, leaves, and stems, but are otherwise identical. Maximal growth rates were determined by allowing a plant with a particular allocation pattern to grow under nutrient- and light-saturated conditions, using the model ALLOCATE. All rates have units of wk^{-1}. Note that allocation to leaf is the overriding determinant of maximal growth rate. Species traits used for these simulations are identical to those used for Figure 4.4–4.6.

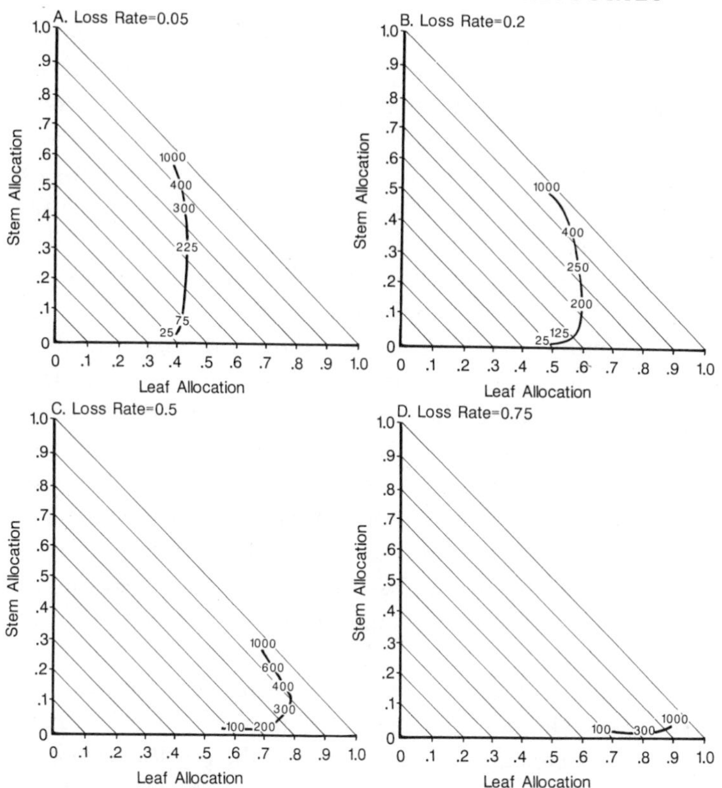

FIGURE 4.8. Competitively superior morphologies for plants that are identical in every way except their allocation patterns. Note that different allocation patterns are favored at different loss rates. Higher loss rates favor plants that have higher leaf allocation, but lower root or stem allocation, because higher loss rates cause increased availability of soil resources and light at the soil surface.

nutrient:light gradient. The analyses were performed by having all 200 species (morphologies) compete for a nutrient and light using the model ALLOCATE. The results (Fig. 4.8) show that different allocation patterns are favored at different loss rates. Higher loss rates shift the competitively superior allocations toward plants with higher allocation to leaves, but lower allocation to roots and stems. The data presented in Figure 4.8 can be viewed in another manner by showing the effect of loss rate on the superior allocation pattern for any given soil type (Fig. 4.9). Higher loss rates favor species that allocate a higher proportion of their biomass to leaves and a lower proportion to roots and stems. This occurs because *higher loss rates lead to a higher average availability of both soil resources and light*, thus decreasing the selective advantage received by individuals that allocate a large proportion of their biomass to roots and stems. There is a continuum from habitats with low loss rates ("undisturbed" habitats) to those with high loss rates. Along this continuum, for any given soil type, the main effect of increasing the loss rate is to favor individuals that allocate more to leaves and less to roots or stems (Fig. 4.9). On poor soils, stem allocation is always low, and individuals are favored that give up root biomass to gain leaf biomass in habitats with higher loss rates (Fig. 4.9). On rich soils, root allocation is low at low loss rates. Higher loss rates favor individuals that give up stem biomass for increased leaf biomass (Fig. 4.9).

For all soil types, increased loss rates favor individuals that have higher maximal growth rates. This is partly because increased loss rates decrease the possible range of viable allocation patterns. For instance, at a loss rate of 0.5 wk^{-1}, only those allocation patterns to the right of the line for which $RGR = 0.5\ wk^{-1}$ in Figure 4.7 are viable. However, the explanation is more complex than that. Habitats with high loss rates have high average availabilities of both nutrients and light, because they have low plant biomass. As discussed in Chapter 3, under conditions of high availability of all resources, the plants with the highest growth rates are likely to be the superior competitors. The far right-hand corner of Figure 4.7 is a plant that allocates nothing to roots or stems, but is all leaf. This plant is basically a soil alga, a green alga, and has the highest relative growth rate of all possible morphologies. It is predicted to be competitively displaced from all habitats except those with extremely high

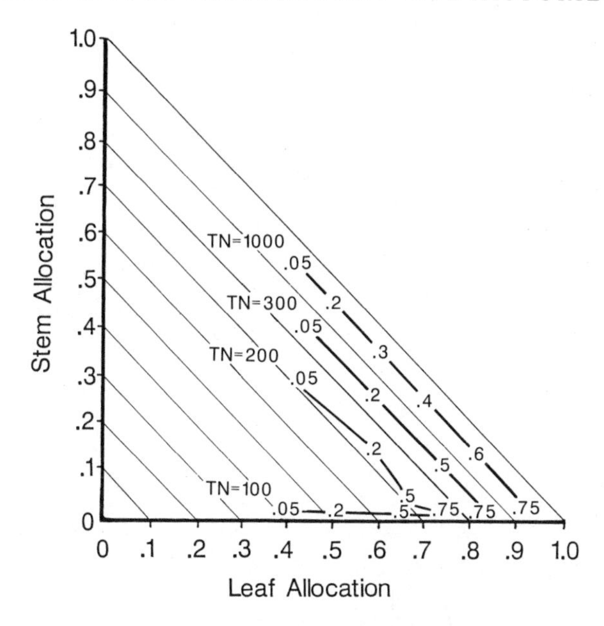

FIGURE 4.9. The dependence of the competitively superior morphology (allocation pattern) on loss rate. Each thick line shows the competitively superior morphologies for a particular soil type (TN). The numbers along each line show the loss rate at which a particular allocation pattern was competitively superior. All results are based on simulations using the model ALLOCATE, with all simulations running for at least 40 years. All parameters, except loss rates, are identical to those used for Figs. 4.4–4.10.

loss rates and rich soils by species with more complex morphologies—morphologies that allow individual plants to obtain more of the resource that limits their growth. The dependence of the competitively superior morphologies on loss rates and nutrient supply rates is shown in Figure 4.10.

Figures 4.9 and 4.10 show that a different morphology is dominant at each unique combination of loss rates and soil nutrient supply rates, and that there is a smooth gradation of morphologies along environmental gradients. Moreover, excluding those morphologies that were inviable because their photosynthesis was always less than losses from respiration and mortality, *each viable morphology is a superior com-*

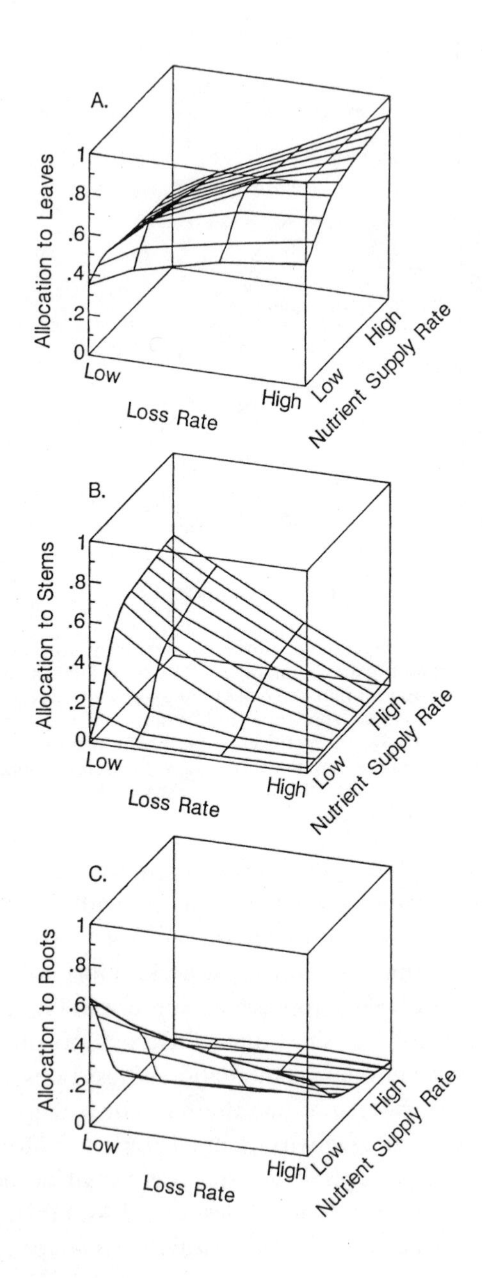

A.
Allocation to Leaves
1
.8
.6
.4
.2
0
Low
High
Loss Rate
Low
High
Nutrient Supply Rate
B.
Allocation to Stems
1
.8
.6
.4
.2
0
Low
High
Loss Rate
Low
High
Nutrient Supply Rate
C.
Allocation to Roots
1
.8
.6
.4
.2
0
Low
High
Loss Rate
Low
High
Nutrient Supply Rate

petitor for a particular habitat type. Because a natural habitat is likely to include a range of nutrient supply rates (i.e., a variety of soil types) and because loss (disturbance, herbivory, etc.) is stochastic and may be spatially patchy (such as treefall gaps or gopher mounds spread over a habitat), the results presented in Figure 4.9 suggest that a range of plant morphologies should be observed in any given habitat. The actual range of morphologies should be determined by the average nutrient supply rate and by plant-to-plant variance in it, and by the average loss rate and by plant-to-plant variance in loss within the habitat mosaic. A given plant community can be considered to fall within some region of the allocation plane. Figure 4.10 thus allows an approximate mapping of the physical characteristics of a habitat into the expected morphology of the dominant species of that habitat.

The work presented here is but a first approximation to the analysis of the relationships between plant morphology, soil nutrient richness and loss rates. There are many other structural, physiological, and environmental constraints that could be included in this model. However, this model demonstrates that a tradeoff faced by all plants constrains an individual plant with a particular morphology and physiology to being a superior competitor for a limited range of habitat conditions.

Physiological Tradeoffs

Just as plants face a tradeoff in their allocation of photosynthate to roots, leaves, or stems, so is it that plants face similar physiological tradeoffs. Physiological processes are enzyme-controlled. Such processes can occur at greater

FIGURE 4.10. The dependence of the competitively superior morphology, as predicted by the model ALLOCATE, on the loss rate and the soil nutrient supply rate of habitats. (A) Dependence of leaf allocation on loss rate and nutrient supply rate. (B) Dependence of stem allocation on loss rate and nutrient supply rate. (C) Dependence of root allocation on loss rate and nutrient supply rate.

rates if enzyme levels are increased. However, greater production of enzymes involved in one physiological process, such as photosynthesis, necessarily means that there are fewer amino acids available for making the enzymes involved in other functions, such as nutrient uptake. The patterns of allocation discussed in this book could be considered simply differences in allocation to various structures such as leaves, roots, or stems. However, even without any differences in leaf, root, or stem mass, a plant could change its physiology so as to modify its competitive ability for a limiting soil resource versus light. Such a change in physiology would also represent a tradeoff, for a plant would be increasing its efficiency at one physiological process by decreasing its efficiency at another. If there were no such tradeoffs, a plant could be a "supercompetitive" species. Directional selection would favor increases in efficiency in all processes until no further increases were possible without a tradeoff. The model developed here, and the tradeoff upon which it is based, is much more general than it might initially seem. Although it did not explicitly include physiological tradeoffs, its behavior is similar to that which would be predicted by a model that did. Holding morphology constant, an individual plant faces a tradeoff in its allocation of protein (nitrogen) to producing efficient photosynthetic mechanisms versus efficient nutrient uptake and retention mechanisms. This perspective is supported by Chapin (1980) who noted that the maximal rates of photosynthesis of a wide variety of plant species were proportional to leaf nitrogen (protein) content.

PLANT HEIGHTS

The results presented in Figures 4.3 to 4.10 showed that average plant heights, as determined by stem allocation, increased with soil nutrient levels. However, in all these cases, the height at maturity of all plants was identical.

Because light tends to be a directional resource, supplied from above, taller individuals intercept more light (Horn 1971; Harper 1977; Grime 1979; Givnish 1982). Slight differences in height can make great differences in the amount of light that a plant captures (Harper 1977). Because light is increasingly less available in more nutrient-rich habitats (at a given loss rate), individuals that allocate more of their potential growth to increases in height and/or leaf area are favored at higher soil resource:light ratios, as already shown (Fig. 4.10). However, allocation to increased height has a structural cost. An upright plant is basically a pole that is anchored at its base. Increased height necessitates increased allocation to below-ground structural tissues used to anchor the pole. Moreover, an upright pole, supported at its base, is structurally unstable (Greenhill 1881; Horn 1971; McMahon 1973; Givnish 1982). Any perturbation away from having a pole point straight upward causes the pole to bend to one side. As it bends, its center of mass is no longer directly over its base, and it thus tends to bend more under its own mass. This moves its center of mass further away from being directly over its base, and bends the pole even more. This bending can cause the pole to buckle. A pole may also fail if it is unable to support its own mass because of compressive stress. For both compressive stress and for buckling, the diameter (D) of a pole should increase as the 3/2 power of its height (H) (McMahon 1973). If a pole is a cylinder, its volume or mass would be proportional to HD^2. For a pole to be stable, though, D should be proportional to $H^{3/2}$. Thus, the mass of a cylindrical pole should be proportional to $H(H^{3/2})^2$, or H^4. This means that an untapered plant at the buckling boundary must increase its stem biomass by a factor of 16 just to double its height. Because plants are tapered, and few plants reach the elastic buckling boundary (Waller 1986), stem volume may increase less rapidly than as the fourth power of height. For the cases considered in this book, I used a less stringent requirement that height was proportional to the square root of stem biomass. Whatever the exact relationship, increased height is energetically expensive, requiring a large allocation to stem tissues.

Considering the energetic cost of increases in height, how should the maximal height at maturity depend on soil nutrient supply rates and loss rates? I used ALLOCATE to explore these ideas, and considered two cases. In both cases, 20 species that were identical in every way except their height at maturity were allowed to compete for a limiting soil nutrient and light. Height at maturity ranged from 5 to 240 cm. In one case, plant mortality was constant ($D = 0.2$), independent of plant height, with all individuals of all sizes experiencing the same density-independent mortality rate. In the second case, there was height-dependent mortality, with all individuals of all morphs having mortality rates that increased linearly with height, from no mortality for plants shorter than 1 cm to a mortality of $D = 0.2$ for plants 250 cm tall.

For the first case, there was neither an advantage nor a disadvantage to individuals of any given height when they competed on poor soils (Fig. 4.11). The final average vegetation height and the final relative abundances of all species were the same as initial values. However, taller individuals were favored on richer soils, with the three tallest morphs displacing all shorter morphs on the richest soils. This suggests that height at maturity can be a neutral trait on soils sufficiently poor that plants are always nutrient limited, but that greater heights at maturity are favored on richer soils on which light becomes limiting. The strength of the selective pressure favoring increased height on richer soils is illustrated by Case 2. In the second case, for which there was height-dependent mortality, short morphs were favored on poor soils, but these were competitively displaced from richer soils by taller, but not the tallest, morphs (Fig. 4.11).

The optimal height at maturity for a plant should be

determined by many factors, including the nutrient and light dependence of its photosynthesis, its morphology, structural constraints that influence its allometry, and the dependence of its mortality on its morphology. The results presented in Figure 4.11 show that height at maturity should increase with the nutrient supply rate. Although this is likely to be a robust prediction, the optimal height at maturity of an individual will also depend on the characteristics of the other plants with which it is competing, making height at maturity have the flavor of a coevolutionary game (Givnish 1982).

SEED SIZE

Seed size is another trait of a plant that should be influenced by the soil-resource:light gradient. In allocating resources to seed production, plants face a tradeoff. The more seeds a plant produces, the smaller the individual seeds must be (Werner and Platt 1976; Harper 1977). If

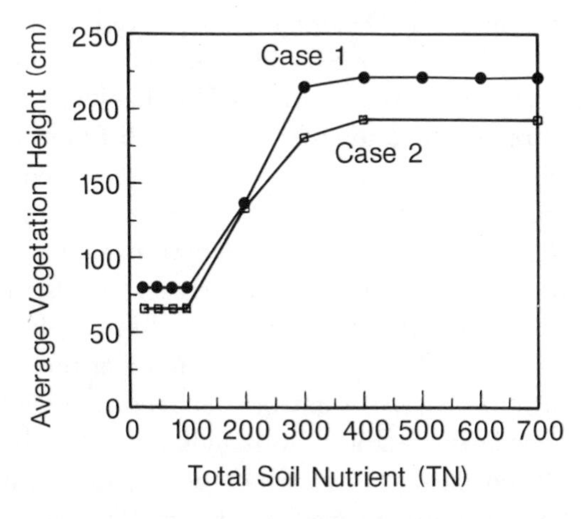

FIGURE 4.11. Average vegetation height (in cm) of the competitively superior morphologies as a function of nutrient supply rate (total soil nutrient, TN). For Case 1, all individuals of all sizes experienced equal mortality rates. For Case 2, taller individuals experienced higher mortality rates.

seed size did not influence survivorship, growth, or fecundity, natural selection would favor the production of many small seeds. Because large seeds allow plants to grow more rapidly as seedlings, and because increases in height, in general, are associated with exponential increases in light, it would seem that large seed size should be favored in habitats in which seedlings are light limited. However, there is another potential cost to large seed. Seeds are a highly nutritious food and thus are subject to predation. Larger seeds are likely to be more easily found, and thus subject to higher predation rates. Can selection for increased seed size be great enough to overcome the costs of seed predation?

I used ALLOCATE to explore two different cases of competition for nutrients and light. In both cases, all 20 competing morphs were identical in all their traits except their seed size. The seed sizes used ranged from 0.001 to 0.090 gm/seed. In the first case, seeds were not subject to any mortality. In the second case, seed mortality was directly proportional to seed size, with 40% of the largest seed dying before germination. Germination occurred immediately after the seeds were produced. With no seed mortality (Case 1), seed size was a neutral trait in habitats with poor soils. The relative abundances of the 20 different morphs did not change through time. Seed size continued to be a neutral trait until TN = 140 (Fig. 4.12). At this point, the individuals that produced larger seed were favored. By TN = 160, the morph producing the largest seed displaced all others. This switch from seed size being selectively neutral to larger seed size being favored occurred at the nutrient supply rate at which newly germinated seeds were light limited. As soon as the seedlings were light limited, individuals that produced larger seeds, which germinated into taller plants, were strongly favored. By TN = 160, all seedlings were light limited, and the tallest, derived from the largest seed, displaced all others. A similar pattern occurred even when seed mortality rates increased with seed size (Case 2,

Fig. 4.12), except that the smallest seeds were favored in the nutrient-poor habitats. Even though the morph producing the largest seed had seed mortality rates that were 90 times those of the smallest seeded morph, and these plants produced only 1/90th as many seeds because of their larger seed size, by TN = 200 the largest seeded morph displaced all other species. Thus seed size should increase along a soil-resource:light gradient. In the absence of seed-size-dependent predation, plants should evolve a seed size that allows seedlings to be sufficiently tall that they are equally limited by nutrients and light upon germination. With size-dependent predation, smaller seed sizes are favored, in general, but seed size still is predicted to increase with soil nutrient supply rate. Small seed size is often discussed as an adaptation that allows plants rapidly to colonize newly disturbed areas (e.g., Werner and Platt 1976). Because disturbance removes plant biomass and thus increases light penetration to the soil surface, the advantage of smaller seeds for

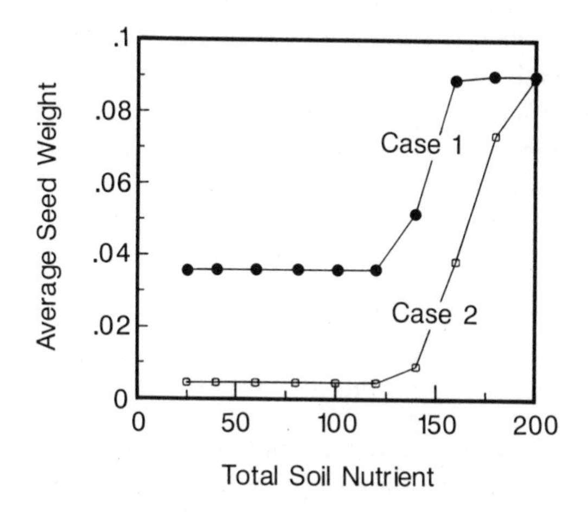

FIGURE 4.12. Average mass per seed of the competitively superior species as a function of nutrient supply rate (TN). In Case 1, all seeds, independent of their size, experienced the same mortality rates. For Case 2, seed mortality rate increased linearly with seed size.

plants living in frequently disturbed habitats may not be just that such seeds increase the chance of migration into such habitats, but also that larger seeds are not favored because of high light availability.

These simulations also illustrate that individuals that produce a few large seeds can be just as successful as individuals that produce many small seeds. This occurs because plants that start from smaller seeds are themselves smaller at all times during vegetative growth than plants that start from larger seeds, at least within the realm of this model. Although plants producing larger seeds necessarily produce fewer seeds, their offspring reach the size of maturity sooner than those starting as small seeds. Within the context of this model of competition among size-structured individuals, seed size is selectively neutral in nutrient-poor habitats unless seed size is related to per capita seed mortality rates or colonization rates. Larger seed sizes, though, could be favored in habitats with resource-poor soils if resource availability increased with depth, as does water availability in xeric habitats. Under such conditions, individuals with larger seeds could produce larger root systems that would provide them with a more than proportionate increase in soil resources (Baker 1972). In this case, a soil resource is assumed to have a depth gradient much like the vertical light gradient, and, as for low light habitats, larger seed size would be favored.

AGE AT FIRST REPRODUCTION AND PERENNIALITY

Age at first reproduction can have a profound effect on the fitness of individuals (Cole 1954). A small increase in the age of first reproduction can cause a large decrease in the reproductive rate of an individual. All else being equal, an individual that reproduces earlier in life will be favored by natural selection because of its greater reproductive rate.

All else, though, is hardly ever equal. Within the framework of the model ALLOCATE, age at first reproduction is a selectively neutral trait on nutrient-poor soils. It is selectively neutral because plants can have two different modes of reproduction: via vegetative growth and via seed. Plants that produce seed sooner in life do so by exchanging the potential for future vegetative growth for current seed production. As long as seedlings and adults have identical mortality rates per individual and grow at identical rates per unit biomass in a particular habitat, production of seed and vegetative growth will be functionally identical and selectively neutral.

However, it seems unlikely that there would be habitats in which seedlings and adults would have identical mortality rates and equal relative growth rates. High rates of herbivory, disturbance, and harsh environmental conditions seem more likely to kill adults than seeds because seeds are small and in a dormant state. If such losses are agents of selective mortality, with adults experiencing higher loss rates than seeds, then increased loss rates would favor annual or short-lived plants over perennials (Charnov and Schaffer 1973).

At any fixed loss rate, the rate of supply of a limiting soil resource could also influence the age at first reproduction. To see this, consider how annual and perennial growth forms within the same species might compete with each other. An annual plant, near the end of the growing season, allocates all its current growth and reallocates all mobilizable compounds to the production of seed (Mooney 1972). A perennial plant, however, must allocate some of its current growth to perennating tissues and to storage, and thus produces fewer seeds during its first year of growth. At the start of the next growing season, all annuals and all seed from perennials would begin growth as seedlings but an established perennial would begin growth with its established root system, meristems, and energy and nutrient stores. The perennial plant would thus be able to support a more rapid initial growth of above-ground biomass than the plants growing from seed. The greater the stores of an established perennial, the more rapid would be its initial growth of above-ground biomass. In a habitat with rich soils in which light would quickly become limiting, an established perennial would have a major advantage because slight initial differences in height would lead to great differences in light capture (Black 1958, 1960; Harper 1977; Newberry and Newman 1978). Annual morphs growing from seed would be competitively displaced by perennial morphs from areas with rich soils because of the early-season height advantage of the perennials in the second and subsequent seasons. In a seasonal environment, the perennating parts of plants function as large seeds. Just as large seeds were favored in more nutrient-rich habitats, so, too, should perenniality be favored in such habitats. Another force favors perenniality in resource-rich habitats, and that is thinning (Harper 1977). As discussed by Westoby (1984), shorter plants are much more likely to die ("be thinned") than taller plants. Thus, on resource-rich soils, the annual growth form is likely to lead to lower early-season height growth and higher mortality rates than the perennial mode, giving a strong selective advantage to perenniality on such soils.

Although perenniality is favored, all else being equal, in nutrient-rich habitats, the annual and perennial growth forms are likely to be selectively neutral on resource-poor soils, if there is no age-dependent differential mortality. One possible source of age-dependent mortality is seasonality. Seasonality is, to some extent, a disturbance that favors dormant structures such as seeds and perennating roots and stems over leaves. If seasonality is extreme enough, it could favor seeds over perennating structures, and thus have annuals be favored in resource-poor habitats. In total, this suggests that the proportion of annuals should be greatest in nutrient-poor, seasonal habitats and lowest in habitats with undisturbed nutrient-rich soils. It further suggests that, if herbivory, disturbance, or other sources of loss cause higher mortality for adults than for seeds, then annual plant abundance should increase with loss rates. There are, though, other selective forces that could favor perenniality, including the possibilities (1) that perennials in a highly seasonal habitat can store resources (Mooney 1972) and thus grow in response to average resource availability whereas annuals would respond to variance in resource availability (Tilman 1982, Figs. 89–90) or (2) that established perennials could better exploit soil resource pulses that occur early in the growing season.

The arguments developed verbally here are qualitatively similar to those developed mathematically for plants that produced seeds of different sizes. In the absence of size-dependent seed mortality, and in a constant habitat, seed size was a neutral trait on soils that were sufficiently poor that seedlings were not light limited. However, on soils that were sufficiently rich that seedlings were light limited, plants that produced larger seeds, and thus taller seedlings, were favored. Perennating structures in plants are analogous to large seeds. The nutrients and energy stored in the perennating structures at the end of one growing season allow a plant to grow rapidly at the start of the next growing season. Such rapid growth is most important in high-nutrient habitats in which light quickly becomes limiting.

As recognized by Raunkiaer (1934), another aspect of perennial morphology, the position of buds, is also important (Fig. 4.13). Annuals and herbaceous perennials, classified by Raunkiaer as cryptophytes or geophytes, have their perennating structures at or below the soil surface. Woody shrubs, classified as chamaephytes, have their buds aboveground on woody stems. Trees, classified as phanerophytes, have buds located at greater distances above the ground. The higher the buds of a plant are above the soil surface, the less likely is it that another individual will shade that plant at the start of the next growing season. This sug-

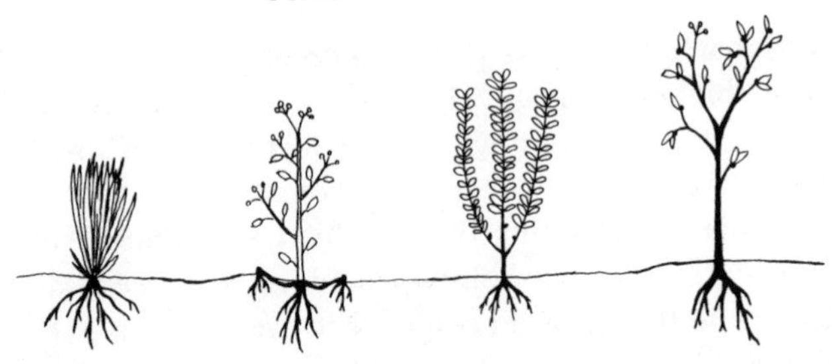

FIGURE 4.13. Location of perennating plant parts, as classified by Raunkiaer (1934). The solid black shows perennating structures, while the open (unshaded) structures are lost each season. The location of leaf buds on perennating structures could be an important component of competitive ability for light in a seasonal environment.

gests that, in seasonal habitats, plants that produce permanent, woody structures that maintain buds above the soil surface are favored in areas with richer soils. However, such woody structures bear several costs, including winter herbivory and the cost of the photosynthate and nutrients required to construct and maintain them. These costs mean that they would not be favored on poor soils. Thus, there should be a fairly smooth gradation along a gradient from poor to rich soils, from geophytes to chamaephytes to phanerophytes, i.e., from short to tall plants, and from plants with low buds to those with high buds.

SUMMARY

The theory presented in this chapter has shown that the pattern of plant allocation to roots, leaves, and stems can greatly influence the maximal growth rates and competitive abilities of plants. For plants living in an unlimited environment, the greatest maximal rates of vegetative growth are achieved by individuals that allocate all their production to additional photosynthetic tissues. Maximal growth rates decrease with the proportion of production that is allocated to non-photosynthetic structures. Despite their decreased rate of growth in unlimited habitats, plants that allocate a proportion of their production to roots and stems are favored in resource-limited environments. For plants living in a particular habitat, characterized by a fixed rate of nutrient supply and a fixed loss rate, there is one particular morphology that leads to superior competitive ability among individuals that are otherwise identical in their physiology and structural constraints. However, each habitat favors a different pattern of allocation to roots, leaves, and stems, i.e., a different morphology.

Both the loss rate of a habitat and its position along a soil-resource:light gradient are important determinants of the superior morphology. If loss rate is held constant, the superior competitors on nutrient-poor soils allocate most of their production to roots and leaves, and little to stems. On richer soils, increased allocation to leaves and decreased allocation to roots are favored. On still richer soils, a large increase in stem allocation is favored mainly at the expense of allocation to roots. On even richer soils, further allocation to stems is favored at the expense of roots and leaves. Thus, along a gradient from nutrient-poor but high light habitats to nutrient-rich but low light habitats, stem allocation increases, root allocation decreases, and leaf allocation is predicted to be maximal at intermediate rates of nutrient supply. Such allocation patterns would cause plants dominant on soils of intermediate fertility to have somewhat greater maximal growth rates than plants dominant on very poor soils or on rich soils. Further, seed size, height at maturity, and age at first reproduction should all increase along a gradient from low soil resource but high light habitats to high soil resource but low light habitats. These generalized traits expected for plants dominant at different points along a soil-resource:light gradient are summarized in Figure 4.14.

Life Histories Along a Soil-Resource:Light Gradient

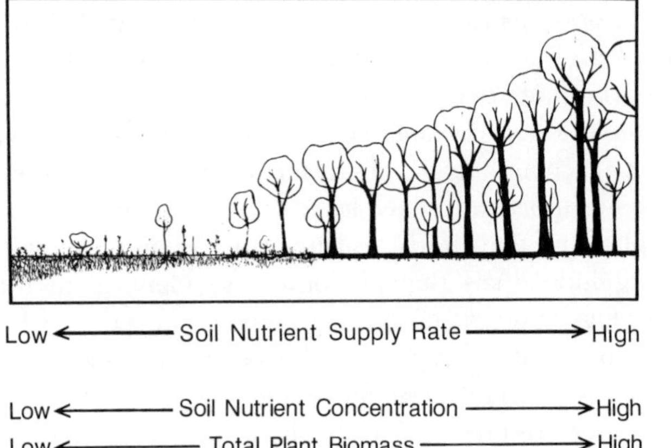

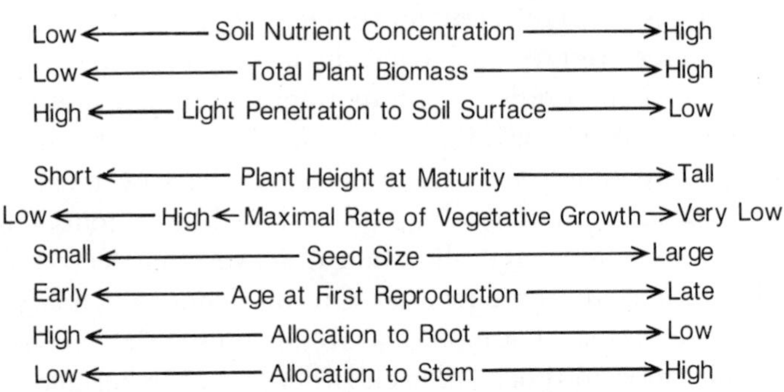

FIGURE 4.14. Predicted dependence of plant morphologies and life histories on productivity for habitats in which loss rate is constant. Predictions are based on simulations using ALLOCATE and other factors discussed in the text. The gradient shown is a gradient from areas with poor soils (low rates of nutrient supply) but high penetration of light to the soil surface to areas with high rates of nutrient supply and low penetration of light to the soil surface.

The physiological traits of plants should also change along productivity gradients. For instance, on nutrient-poor soils, there would be strong selection favoring individuals that required lower tissue nutrient levels, even if this decreased the light-saturated rate of photosynthesis, as Chapin (1980) suggests may occur. On nutrient-rich soils, individuals with higher light-saturated rates of photosynthesis and with lower light compensation points would be favored. Morphological, physiological, and life history traits all interact to determine the ability of a plant to compete. There should be a particular suite of such traits that allows a plant to be a superior competitor for a particular habitat. A different suite of traits would be favored in a habitat with a different nutrient supply rate or different loss rate. Although I have discussed the evolution of plant morphologies as if plant physiologies were fixed, morphology, physiology, and life history should be interrelated and interdependent. All should change in concert along productivity and loss rate gradients.

Loss rate (from herbivory, disturbance, etc.) also has a large effect on the traits of the plants that are superior competitors on particular soils. On rich soils, the plants dominant at low loss rates are taller, have lower maximal rates of vegetative growth, are more long-lived and produce seed later in life, and have higher allocation to stem and lower allocation to leaves compared to the plants that are favored by high loss rates (Fig. 4.15). Root allocation is fairly constant along this gradient. For a loss rate gradient on poor soils, plant stature is not highly dependent on loss rate, but the plants dominant at low loss rates allocate more to roots and less to leaves than those dominant at high loss rates. This greater allocation to leaves in habitats with higher loss rates, which occurs on both poor and rich sites, means that plants dominant at higher loss rates have greater maximal growth rates (RGR_{max}). These higher maximal growth rates occur because higher loss rates increase the average availability of both nutrients and light by decreasing average total plant biomass. Higher levels of all limiting resources favor individuals that have higher growth rates.

These major differences between plant traits favored along loss rate gradients versus those favored along productivity gradients come from the different patterns of resource availabilities along these gradients. Holding loss rate constant, the concentration of a limiting soil resource is negatively correlated with light penetration to the soil sur-

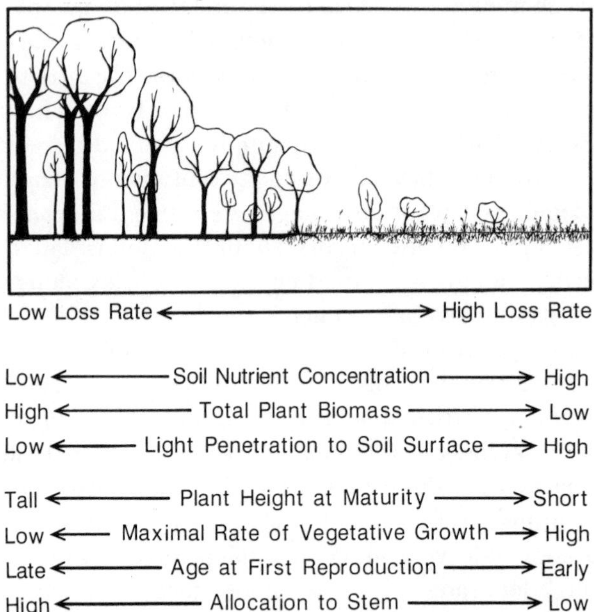

FIGURE 4.15. Predicted dependence of plant morphologies and life histories along a loss rate gradient on which nutrient supply rate is constant. Predictions are based on simulations using the model ALLOCATE. The gradient is from areas with low loss rates to areas with high loss rates. The average availability of the limiting soil resource and of light at the soil surface increases with loss rate.

face along a productivity gradient. In contrast, nutrient availability and light penetration to the soil surface both increase (i.e., are positively correlated) along a gradient from low to high loss rates. Because the morphology of a plant determines the ambient levels of the soil resource and of light at the soil surface (or at some other height or heights) for which a plant will be a superior competitor, these different resource gradients lead to qualitatively different patterns of species replacements. Clearly, a productivity gradient or a loss rate gradient are just two of an unlimited number of possible trajectories in the full nutrient supply rate and loss rate plane (Fig. 4.10). Each trajectory will lead to a different sequence of plant morphologies and life histories. Both the rate of supply of the limiting soil nutrient and the loss rate influence the morphological traits that lead to superior competitive ability in a particular habitat, and thus both are important, inseparable elements influencing the vegetational composition of a habitat.

CHAPTER FIVE

Vegetation Patterns on Productivity and Loss Rate Gradients

As discussed in Chapter 4, many aspects of plant morphology and life history are expected to depend on the nutrient supply rate and the loss rate of a habitat. For instance, plant species are predicted to be separated, in an individualistic manner, along major environmental gradients, with a correspondence between the point along such gradients at which a species reaches its peak abundance and its morphology, life history, and physiology. Plants dominant in habitats with lower loss rates are expected to have lower maximal relative growth rates than those dominating habitats with higher loss rates. Seed size and allocation to stem are predicted to increase along a productivity gradient. Closed-canopy habitats are predicted to have higher rates of supply of a limiting soil resource and be dominated by taller plants than open-canopy habitats. How well are such predictions supported by available observational and experimental studies? I will first consider the predicted dependence of maximal growth rates on habitat loss rates and nutrient supply rates. Then I will consider several other broad-scale patterns within and among plant communities.

MAXIMAL GROWTH RATES

The maximal rate of vegetative growth of a plant should increase with the proportion of its production that is allocated to additional photosynthetic tissues (Chapter 3 this volume; Monsi 1968). Thus, non-nitrogen-fixing soil algae should have the most rapid maximal growth rates, and plants with the greatest allocations to roots and stems should have the lowest maximal growth rates. ALLOCATE predicted that the proportion of photosynthate allocated to leaves, and thus maximal growth rate, would depend on both the loss rate and the nutrient supply rate of a habitat (see Figs. 4.7 and 4.8). Figure 5.1 shows the predicted dependence on habitat loss rate and habitat nutrient supply rate for the maximal growth rates of the competitively superior morphologies. Holding nutrient supply rate constant, RGR_{max} should increase with loss rate. Holding loss rate constant, RGR_{max} should be maximal in habitats of intermediate productivity for low to moderate loss rate habitats, and increase with productivity for high loss rate habitats. In general, habitat loss rates are predicted to be a more important determinant of the RGR_{max} of the dominant species than habitat nutrient supply rates. The difference between the fastest growing and the slowest growing plants along a productivity gradient is small compared to the difference along a gradient from low to high loss rates.

These predictions are based on the form of ALLOCATE used in Chapter 4. A realistic modification to a parameter used in those simulations might lead to a more rapid and dramatic decline in RGR_{max} for habitats beyond the peak reached at intermediate rates of nutrient supply along a productivity gradient. This is because the allometric relation used to describe the dependence of plant height on stem biomass assumed that plant height, H, was related to stem biomass, S, as

$$H = cS^{0.5}, \qquad (5.1)$$

i.e., that stem biomass had to quadruple for height to double. I chose this relationship because it approximated relations I have found for several grass species. However, as

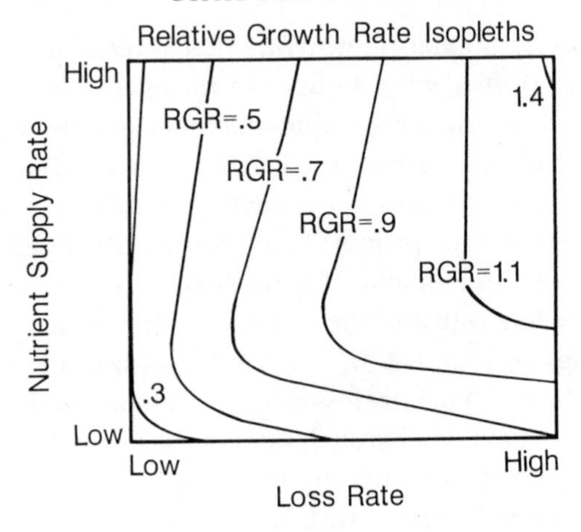

FIGURE 5.1. The dependence of the maximal relative growth rate (RGR) of the competitively superior species on the loss rate and the nutrient supply rate of a habitat. The curves in the figure are RGR_{max} isopleths, i.e., lines of equal maximal relative growth rates. Increases in loss rates favor species with higher RGR_{max}. Holding loss rate constant, RGR_{max} is predicted to be a "humped" function of the nutrient supply rate of a habitat, for habitats with low to moderate loss rates. At high loss rates, RGR_{max} is predicted to increase with nutrient supply rate. The isopleths shown here are based on the simulations of ALLOCATE presented in Figures 4.7, 4.8, and 4.10.

discussed in Chapter 4, for a pole to be stable, its height should increase as the fourth root of its stem mass (McMahon 1973):

$$H = cS^{0.25}, \tag{5.2}$$

i.e., stem mass would have to increase 16-fold for height to double. To consider plants ranging in height from mosses to sequoias, then, a much greater allocation to stem at greater heights may be required than was assumed. Such greater allocation to stem, though, must come at a cost in allocation to leaves and roots. This means that the expected maximal growth rates of species dominant on very nutrient-rich soils should be quite low, and might fall more rapidly along a productivity gradient than illustrated in Figure 5.1.

Both Grime (1979) and Chapin (1980) offered empirical evidence that plants common in very infertile habitats have low maximal relative growth rates. Both stated that plants dominant in fertile habitats have high maximal relative growth rates. Unfortunately, the latter data could be interpreted as being consistent or inconsistent with the patterns predicted for productivity gradients, depending on whether or not the habitats Grime and Chapin called "fertile" were at or beyond the peak in leaf allocation that occurs in habitats of intermediate fertility. If, though, Grime and Chapin are correct that the species dominant on the most fertile sites do, indeed, have the largest RGR_{max}, then their data would contradict a major prediction of this theory. Further, there are many significant differences between the view developed in this book and the view espoused by Grime (1979). I will take this opportunity to explore the empirical evidence in greater depth and to compare these divergent approaches to plant evolution and community structure.

Grime (1979) drew his data from a study of the maximum relative growth rates of seedlings of 132 species of flowering plants of the Sheffield, England, region (Grime and Hunt 1975) and from his perceptions of the fertility of the habitats in which each of these species is now found. As discussed in Chapter 3, Grime and Hunt (1975) found that annuals had the greatest maximal relative growth rates, that herbaceous perennials had lower rates, and that woody perennials had the lowest maximal relative growth rates (see Fig. 3.2). What, though, is the relationship between the maximal relative growth rates of these 132 species and the fertility of the habitats in which each naturally occurs?

In order to determine the relationship between RGR and habitat fertility, it is necessary to know how the natural abundances of these plant groups depend on soil fertility. As emphasized repeatedly by Tansley (1949), however, this is not an easy task because of the great changes in the British flora that have occurred due to deforestation and heavy grazing. Tansley (1949, p. 129) stated: "Originally derived from forest land, the pasture or meadow cannot be recolonized by woody plants because of the continued grazing; and we have here a plant community conditioned on the one hand by climate and soil and on the other by the constant factor of grazing." In many areas that are currently pasture, but which were forest before being cleared by humans, *all that is required for the reestablishment of trees is the erection of fences that exclude mammalian herbivores* (Farrow 1916, 1917; Summerhayes 1941; Tansley 1949).

What, then, was the vegetation like before clearcutting and heavy grazing? The most infertile sandy soils may have had a heath vegetation dominated by ericaceous shrubs (Tansley 1949) or the heath may have occurred only in small pockets or barrens within scrub or woodland on these sandy soils. Tansley stated that herbaceous pasture grasses and forbs were, before Neolithic and modern grazing, found on the areas free of woody vegetation, specifically "on the poorer sands, on recently formed terraces of alluvial river gravel, . . . and also on hill tops and wind-swept crests such as the narrower ridges of the chalk downs, where the soil is very shallow" (Tansley 1949, p. 164). In contrast, forest covered the areas with deeper or richer clay or loamy soils (Tansley 1949, p. 163; Godwin 1956).

Thus, before cutting and grazing, the most fertile areas in the British Isles were dominated by trees. Of all the plant groups Grime and Hunt (1975; Fig. 3.2 this volume) considered, woody perennials have the lowest RGR_{max}. Indeed, of the 132 plant species studied by Grime and Hunt, 5 of the 6 most slowly growing species were trees (*Acer pseudoplatanus, Picea abies, P. nigra, P. sitchensis, P. sylvestris*). Trees, the dominant life form on very fertile soils, have the lowest RGR_{max} of all plant life forms. The group with the next lowest maximal relative growth rates, based on Grime and Hunt's data, are woody perennials common to heath. *Vaccinium vitis-idaea*, the cowberry or lignon berry, had the second lowest maximal growth rate of all 132 species, and two other heath species (*Calluna vulgaris* and *Vaccinium myrtillus*) were among the 10 most slowly growing species. Thus, plants common in one of the most nutrient-poor habitats, heath, and plants dominant in the most fertile habitats, forests, have the lowest growth rates of all species studied by Grime and Hunt (1975). Legumes, which also reach their peak abundance on very nutrient-poor soils (Tilman 1982), also had very low RGR_{max}. Grasses, forbs, and woody shrubs, dominant in areas with soils of intermediate fertility before clearcutting and agricultural grazing, have much higher maximal relative growth rates. When the data gathered by Grime and Hunt are interpreted in terms of the habitats in which species were common before the major human disturbance of deforestation and the imposition of grazing losses, these data support the predictions of Chapter 4, but completely contradict the frequently cited generalization made by Grime (1979) that plants dominant in fertile habitats have the highest maximal growth rates.

Why do Grime and Hunt's own data seem to contradict their conclusions? The problem comes from their reliance on the present distributions of these species. The fertile lands that were forested before Neolithic times (Godwin 1956) have now become the most productive fields and pastures of England. Because pastures are stocked with sheep or cattle to the greatest herbivore density that the fields can support, more fertile fields are more heavily stocked. The relationships reported by Grime and Hunt between present-day vegetational composition and soil fertility are thus confounded by the correlation between soil fertility and herbivore stocking density. Further, the bias induced by clearcutting and heavy grazing cannot be eliminated by a few years free from grazing. Woody plants are slow-growing, immensely large at maturity, and currently rare in these regions. Decades, if not a century or more, are required to re-establish the natural (i.e., pre-disturbance) correlations between soil nutrient levels and plant traits. Further, as will be discussed in Chapter 6, differences in maximal growth rates could lead to decades during which the observed dynamics of various plant species are the opposite of the long-term pattern that will eventually develop. Such transient population dynamics greatly add to the difficulty of interpreting relationships between soil fertility and plant distributions, especially in successional vegetation on rich soils (Tilman 1985; Chapter 6 this volume).

Grime (1979) is correct that the most fertile areas, arable fields, are *currently* dominated by the most rapidly growing herbaceous perennials and annuals. Although Grime and Hunt (1975) tried to allow for the impact of grazers on rich soils by eliminating annual plants from their analyses, they did not realize the magnitude of the grazer effects. Even the pastures that are not currently grazed by sheep are still likely to have what Tansley (1949) asserted were unusually high grazing rates caused by the great increases in hare, rabbit, and small mammal densities associated with the decimation of their natural predators, as well as the problem of transient population dynamics. Because these arable pastures were forested until they were cleared and grazed (Tansley 1949; Godwin 1956), and because forest often re-establishes itself if herbivores are excluded, the presence of rapidly growing species in these fertile soils is best ascribed to the combined effects of heavy grazing losses, grazer-caused disturbances, and soil fertility. As Tansley's and Godwin's descriptions of the original vegetation of England clearly demonstrate, the present distributions cannot be ascribed just to the high fertility of the soils.

I interpret Grime and Hunt (1975) and Grime (1979) as having studied a gradient from poor soils that had low loss rates from grazing and disturbance to rich soils that had high loss rates. The areas with the lowest loss rates were dominated by plant species with the lowest RGR_{max} whereas the areas with the highest loss rates were dominated by species with the highest RGR_{max}. When viewed in this manner, their conclusions are completely consistent with the predictions of Chapter 4. A gradient from low loss rate, nutrient-poor habitats to high loss rate, nutrient-rich habitats is predicted to lead to an almost linear increase in RGR_{max}, as can be seen in Figure 5.1.

The stark contrast between the vegetation of grazed pastures and the vegetation of such areas before clearcutting and grazing illustrates the role of herbivory and other sources of loss or disturbance in structuring and maintaining these communities. On rich soils, areas with natural herbivore densities had closed-canopy forests. The high loss rates imposed by livestock, though, has led to a pasture vegetation dominated by short, fast-growing species. The natural, wooded vegetation (which returns, albeit slowly, when grazing is halted) and the pastures are thus at opposite ends of a loss rate gradient. Vegetation structure, plant life histories, and maximal growth rates along this gradient are consistent with the predictions in Figure 4.15.

However, there is an equally important lesson to be learned from the contrast between grazed and ungrazed areas. The herbivores that maintain these areas as pastures are stocked herbivores that are kept at unusually high densities because they are protected from predators, parasites, and diseases. According to Tansley (1949), before the natural predators were decimated in these areas, predators kept natural herbivores at much lower densities. At those natural densities, herbivores were not able to prevent the establishment and maintenance of forest in localities with rich soils. Thus, it is not the presence of herbivores, per se, that confuses the issue, but the presence of unusually high densities of herbivores caused by predator control or predator decimation. This suggests that the structure of these communities, and potentially all natural communities, may depend on multi-trophic-level interactions.

There is much to learn about multi-trophic-level interactions, and, unfortunately, little time to learn it because of the destruction of natural habitats and the continued decimation of predators in habitats around the world. Each species lives within the context of the entire food web. There is currently no reason to believe that herbivores will always be held by their predators and parasites at densities at which they do not greatly affect plants. Nor is there yet a clear empirical, experimental, or theoretical foundation upon which to base any firm conclusions about the relative role of herbivory and predation in communities that differ in their productivity. Two pioneering attempts to address this issue were those of Hairston, Smith, and Slobodkin (1960) and of Oksanen et al. (1981). Hairston, Smith, and Slobodkin suggested that many plants may be limited by their resources and that many herbivores may be limited by their predators and parasites. They suggested that competition was probably the main biotic interaction determining the structure of plant communities, and that predation was the main interaction structuring herbivore communities. I disagree with their simplification that an organism is either limited by its resources or limited by the organisms it consumes, because, as discussed in Chapter 2, resource-dependent growth and mortality or loss must always interact to control population sizes. However, their assertions can be rephrased to state that predators often keep herbivores at much lower densities than would occur if the predators were absent. The lower herbivore densities cause lower loss rates and disturbance rates for plants, and thus favor more slowly growing plant species that are efficient at growing at low resource availabilities. When interpreted in this manner, many of the examples cited by Tansley (1949) seem to support their claim, especially when compared with the patterns in British vegetation that have followed predator extermination.

These ideas were re-explored and expanded in Oksanen et al. (1981). They suggested that the structure of food webs would change in relation to primary productivity. In habitats with very resource-poor soils, they argued, herbivore densities would be held to very low levels because of the low quantity and quality of the food available for them. More productive habitats would be able to support a proportionately greater density of herbivores because of greater plant biomass, but could still have sufficiently low herbivore densities that predators would be rare. The most productive habitats, though, would have sufficient herbivore densities to support high densities of their predators, and the proportionate effect of herbivores on plants would be low. This last case, they suggest, is the case described by Hairston, Smith, and Slobodkin (1960). Some of the possible implications of such multi-trophic-level interactions are discussed in Chapter 9.

These patterns demonstrate the potential danger of ignoring the full range of trophic relations in an ecosystem. The lesson to be learned from this reanalysis of the relation between soil fertility and growth rates is that soils can affect the structure of plant communities both directly as well as indirectly through correlations between fertility and herbivory. Correlational studies in natural habitats are clearly unable to determine what the underlying causes of any given pattern might be. A rigorous experimental approach in which nutrients, herbivores, and predators are experimentally varied is clearly required to determine what factors may be causing any given pattern in nature. All organisms live embedded within a food chain. All species are consumers and all are resources for predators, parasites, herbivores or decomposers. A mechanistic approach, such as that developed in this book, which details the ways in which organisms interact with their resources, can be easily expanded to include other trophic interactions. With this approach, it is possible to build a mechanistic and potentially predictive understanding of the forces structuring the world's biota.

PLANT PHYSIOGNOMY

One of the most universal observations that has been made in vegetation around the world is that the average height at maturity of a species tends to correspond with the fertility of the habitat in which that species reaches its peak relative abundance (Beard 1944, 1955, 1983; Whittaker 1975; Mooney 1977). The general importance of this pattern has been overlooked, perhaps because its similarity to variation within individual species along fertility gradients makes it seem trivial. However, there is a difference between an individual plant growing taller following nutrient addition and species that can be taller at maturity displacing shorter species on richer soils. The former might be dismissed as a morphological necessity; the latter requires an ecological explanation. Even the former, however, is not an absolute ecological necessity. Though increased nutrient availability can cause plants to grow more rapidly, why should such growth be allocated so as to make a plant taller? Light is the resource that becomes more limiting as soil nutrient levels increase on a productivity gradient. Increased height can be explained best as an adaptation that allows individuals to acquire more light. Similarly, the tendency for shorter species to be displaced by taller species along productivity gradients is likely to be caused by taller plants being better competitors for light.

Let us first consider plant height along natural productivity gradients. For instance, in the regions of California, Chile, South Africa, and the Mediterranean with mediterranean climates, there has been a pronounced convergence of many morphological and physiological features of phylogenetically unrelated plant species (e.g., Mooney 1977; Cody and Mooney 1978). All these areas are dominated by an evergreen scrub vegetation. Along gradients from coarse, well-drained soils to finer-textured soils, the vegetation changes from a sparse scattering of species that are less than 0.5 m tall at maturity to a dense, closed scrub composed of species that are more than 5 m tall. Rich soils that have not been subjected to recent disturbance tend to be dominated by a closed canopy forest with trees 10 m tall. For example, within the Coolgardie Botanical District of southwestern Australia, Beard (1983) found that the vegetation ranged from 3 m tall scrub heath on the sandy soils to 25 m tall salmon gum woodlands on the white, kaolinitic clays, the most productive soils of the district. Taking a broader geographical view of the vegetation of this entire region, Beard (1983) found that, if he held parent material constant, there was the same sequence from 3 m scrub heath to 25 m forest along a climatic gradient from areas receiving 200 mm/year of rainfall to those receiving 1200 mm/year. Thus, there was a strong similarity between small-scale vegetational patterns caused by soil parent material and broad, geographic trends associated with climate.

This similarity may be caused by water being limiting in both cases. Differences in soil type could influence water availability, and thus control local productivity, just as water availability may control productivity along the geographic climatic gradient. Alternatively, a different soil resource, such as nitrogen or phosphorus, may limit growth on all geographic scales. The correlation with water may come because water may influence the rates of accumulation and microbial degradation of organic matter, and thus influence rates of mineralization of the limiting nutrient. The original parent material may similarly influence the rate of accumulation and/or mineralization of a limiting soil resource within a local habitat. However, whatever the underlying limiting soil resource or resources may be,

Beard's analysis shows that unproductive areas are dominated by a sparse cover of plant species of short stature, whereas increasingly productive areas are dominated by a denser cover of taller species.

In Wisconsin, Michigan, and Minnesota, some of the least productive sites are areas with coarse sandy soils. The most productive sites tend to occur on soils derived from glacial till or former lake sediments. Controlling for fire history and human disturbance, the least productive areas tend to be dominated by soil algae, lichens, mosses, ferns, grasses, and forbs, with scattered pin cherry or choke cherry. Richer sandy soils have an oak savannah vegetation. Still richer soils are dominated by a closed-canopy oak forest. The richest soils have a sugar maple and basswood closed-canopy forest (Hole 1976; Olson 1958; Whitney 1986). The species dominant on poor soils are short at maturity, and vegetation height tends to increase with soil nutrient richness. As in mediterranean areas, the correspondence between soils, including original parent material, and vegetation is quite dramatic in Wisconsin (Hole 1976).

Givnish (1982) reported that the density of understory herbs was greatest in forested areas on poor soils, and lowest in forested areas on rich soils. This occurred because forests on richer soils attenuated more light, and these low light levels decreased the density of the understory herbaceous species. He further reported a pattern that, on its surface, seems contradictory to the generalizations made above. For the understory herbs (but not for the canopy trees), the tallest species were dominant on the poorest soils and the shortest species dominant on the richest soils. However, Givnish asserted that the understory plants were mainly competing with each other for the light remaining after light had been captured by the canopy trees. In this situation, light capture is dependent on height only when the herbaceous plants are at high densities. When there is a closed overstory canopy and little light reaches the soil surface, understory herbs are rare, and there is little chance one herb will be shading another. In these situations, there is little gain in light intensity from increased plant height, and taller individuals would not be favored (Givnish 1982). Thus, Givnish's observations are consistent with the qualitative predictions of theory—that increased plant height is favored if a plant is light limited and if an increase in its height leads to increased light intensity.

Beard (1944, 1955) reported that one of the major patterns of variation in the "climax" vegetation of the New World tropics was determined by annual rainfall (Fig. 5.2A). In comparing areas along a gradient from rainfall one month of the year to sufficient rainfall every month of the year, he observed a smooth intergradation in the species composition of the vegetation (Fig. 5.2A). He termed this the "seasonal formation series." For ease of discussion, Beard divided this continuum into several different formations. The least productive area he studied, desert, was dominated by "little, thorny shrubs, succulents and herbs" (Beard 1944, p. 141). Next along this rainfall gradient was cactus scrub, dominated by columnar cacti and prickly-pears (2–4 m tall), with scattered thorny shrubs. Thorn woodland, a scrubby, somewhat open vegetation dominated by evergreen spiny trees from 3 to 10 m tall, occurred in areas with greater rainfall. This was followed by deciduous seasonal forest dominated by low-branching trees 15 to 20 m tall, with lianas and arboreal epiphytes being rare. Next along this seasonal rainfall gradient came semievergreen seasonal forest, a closed-canopy forest 20 to 26 m tall, with lianas quite abundant but epiphytes rare. Then came evergreen seasonal forest, which had occasional emergent trees reaching 35 m, and an intermediate stratum ranging in height from 14 to 30 m. Epiphytes were well developed and lianas were common. The areas which received sufficient rainfall every month and which had the greatest total annual rainfall contained tropical rainforest, dominated by

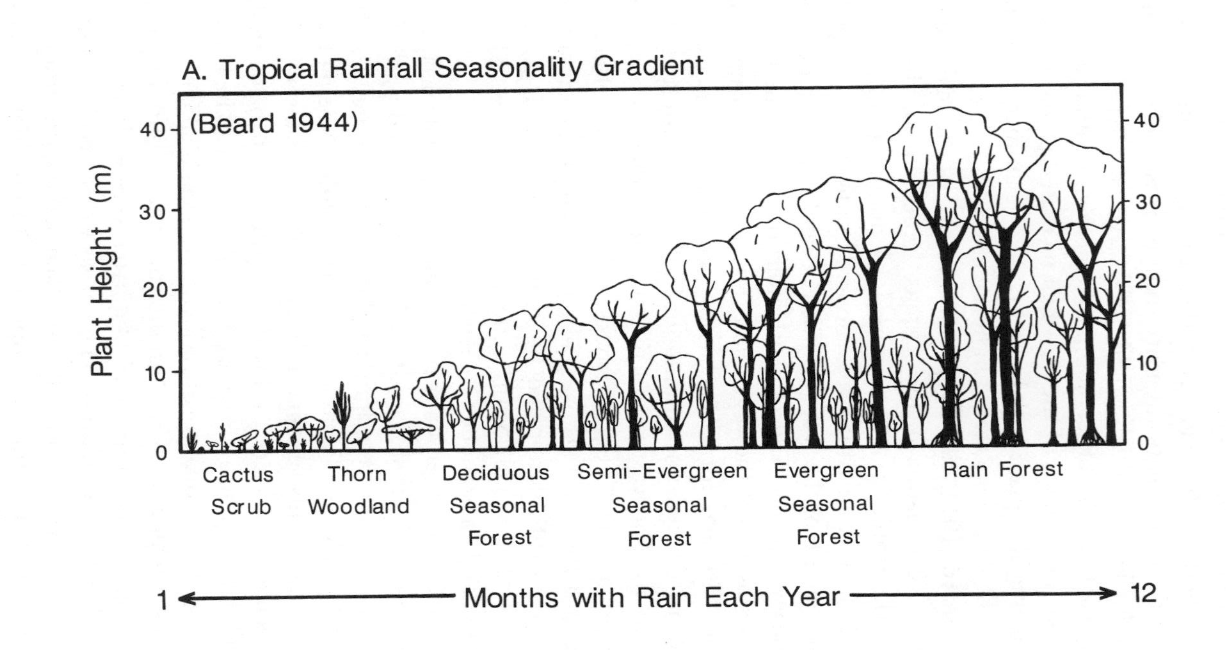
A. Tropical Rainfall Seasonality Gradient
(Beard 1944)
Plant Height (m)
40
30
20
10
0
Cactus Scrub
Thorn Woodland
Deciduous Seasonal Forest
Semi-Evergreen Seasonal Forest
Evergreen Seasonal Forest
Rain Forest
1
Months with Rain Each Year
12

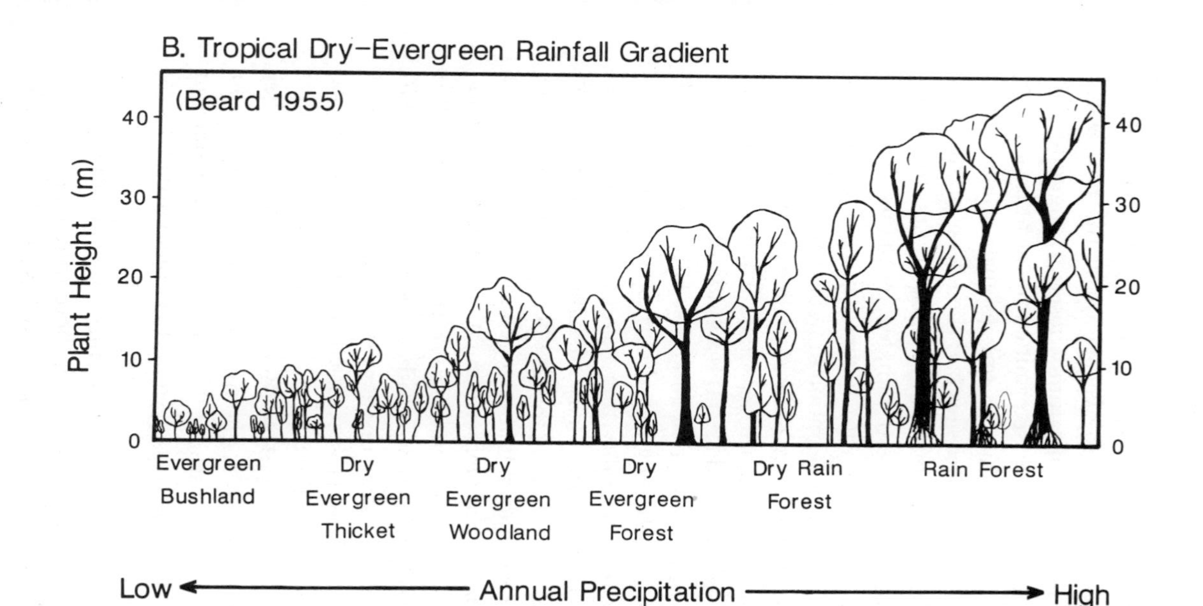

FIGURE 5.2. (A) Plant morphologies along a geographic gradient in rainfall seasonality in the New World tropics. The habitats included range from those with one month of sufficient rainfall per year to those with sufficient rainfall every month. Figure redrawn and modified from Beard (1944). (B) Plant morphology along a gradient in average annual rainfall, but for habitats with rain falling each month of the year. Note that dry areas with rainfall each month have an evergreen brushland vegetation, whereas dry areas with rainfall in just one month per year (A) have a cactus scrub vegetation. Modified from Beard (1955).

an upper stratum of 40 m tall trees and one or two more lower strata. Lianas were rare and epiphytes occurred only near the canopy (Beard 1944). In its broadest view, then, Beard (1944, 1955) saw the vegetation of the new world tropics as forming a continuum whose physiognomy and species composition were determined by, or at least correlated with, the availability of a soil resource, water. Along this continuum, the driest sites had the lowest productivity and the highest proportion of light penetration to the soil surface. These sites were dominated by plants that were short at maturity. Sites with higher rainfalls had higher plant biomass, lower penetration of light to the soil surface, and were dominated by increasingly taller plants.

I have little doubt that the seasonal formation series gradient described by Beard (1944) is a gradient in the relative availabilities of soil resources and light. However, although the patterns described by Beard correspond with the seasonality of water availability, there is no direct experimental evidence that water is actually an important limiting soil resource at any point along this gradient. Forestry practices in tropical areas suggest that nitrogen can often be limiting (Vitousek 1982). Direct experimentation is needed to determine what resources are limiting at various points along this and any other productivity gradient. Such research could also provide information on the response of various species to resource additions. However, whatever soil resources are found to be limiting, the tropical seasonal formation series described by Beard (1944, 1955) clearly demonstrates changes in plant physiognomy along soil-resource:light gradients that are consistent with the theory developed in Chapter 4. Thus, the theory seems to explain broad, trans-biome patterns.

Beard (1944, 1955) also recognized several other environmental gradients along which there were marked separations of the dominant species of the New World tropics (Figs. 5.2, 5.3). All of these gradients are productivity gradients, and each represents a particular environmental variable or suite of variables that can cause productivity to be reduced from that of lowland tropical rainforest. For each gradient, decreases in productivity lead to changes in species composition, with shorter species dominating less productive habitats. The "dry evergreen formation series" occurs in habitats in which water availability is constant throughout the season, i.e., all months have almost equal rainfalls (Fig. 5.2B). The wettest areas are dominated by tropical rainforest and the driest by rock pavement vegetation (Fig. 5.2B). Another productivity gradient recognized by Beard (1944, 1955), caused by elevation and the associated decreases in temperature and growing season length (Fig. 5.3A), was termed the "montane formation series," which ranged from tropical rainforest to tundra. Beard (1955) also recognized two swamp formations, with vegetation composition determined by whether or not the area was flooded all year or only seasonally (Fig. 5.3B) and by the depth of the standing water.

Whittaker (1975) summarized two additional gradients, which he termed ecoclines. The first was a moisture gradient from the desert of southwestern North America to the mesophytic forest of the Appalachian Mountains (Fig. 5.4A). Along this gradient cacti and desert shrubs give way to dry grasslands, then to short grass prairie, tall grass prairie, oak savannah, oak woodlands, oak hickory forest, and finally sugar maple, basswood and/or tulip tree forests. The second was a latitudinal and thus temperature gradient from seasonal tropical forest northward to boreal forest and arctic tundra (Fig. 5.4B).

The six gradients of Figures 5.2 to 5.4 and the other gradients described earlier in this chapter are amazingly similar, even though they represent unrelated species living in geographically distinct habitats. Along all of these gradients, the most productive habitats are dominated by a tall, closed-canopy forest with a high leaf area index. Any envi-

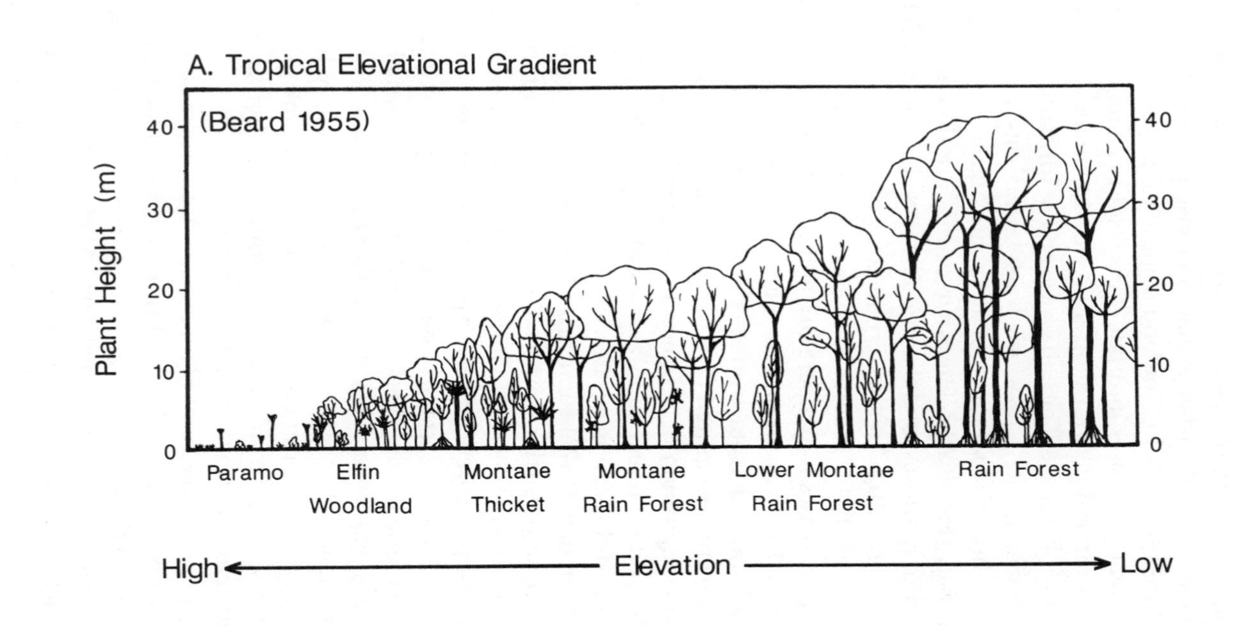
A. Tropical Elevational Gradient
(Beard 1955)
Plant Height (m)
40
30
20
10
0
Paramo
Elfin Woodland
Montane Thicket
Montane Rain Forest
Lower Montane Rain Forest
Rain Forest
High
Elevation
Low

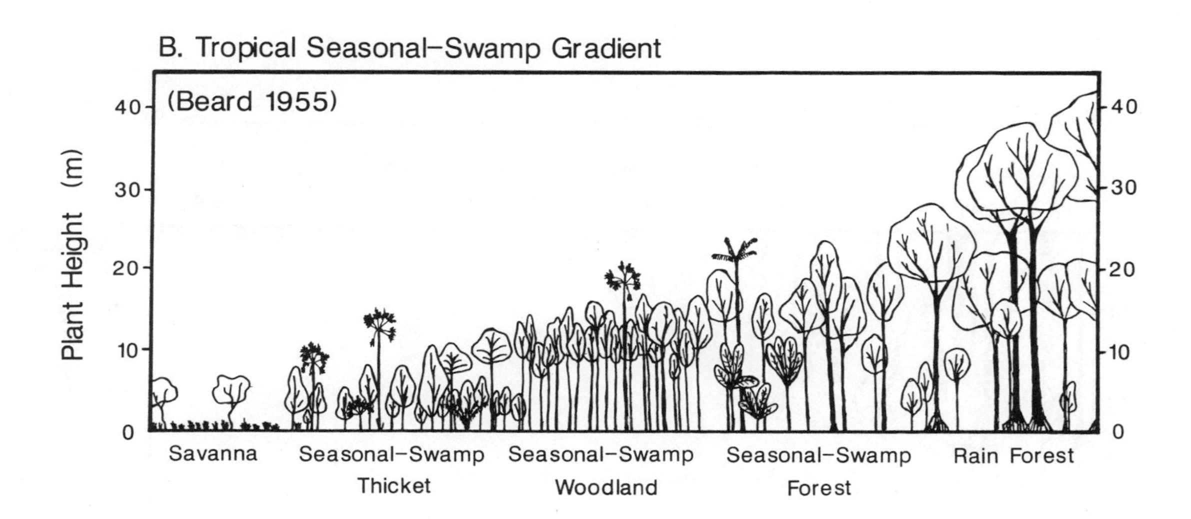

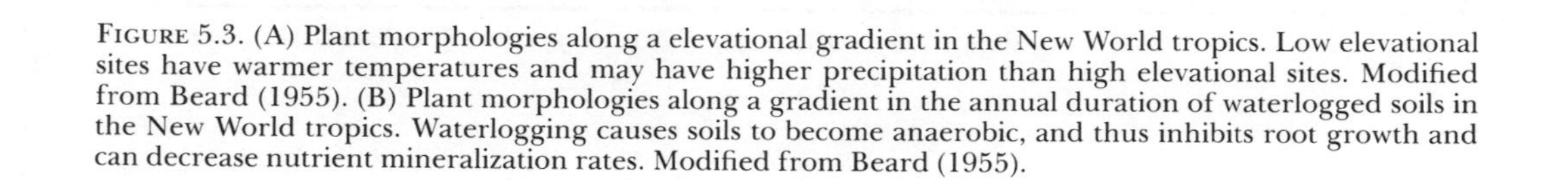

FIGURE 5.3. (A) Plant morphologies along a elevational gradient in the New World tropics. Low elevational sites have warmer temperatures and may have higher precipitation than high elevational sites. Modified from Beard (1955). (B) Plant morphologies along a gradient in the annual duration of waterlogged soils in the New World tropics. Waterlogging causes soils to become anaerobic, and thus inhibits root growth and can decrease nutrient mineralization rates. Modified from Beard (1955).

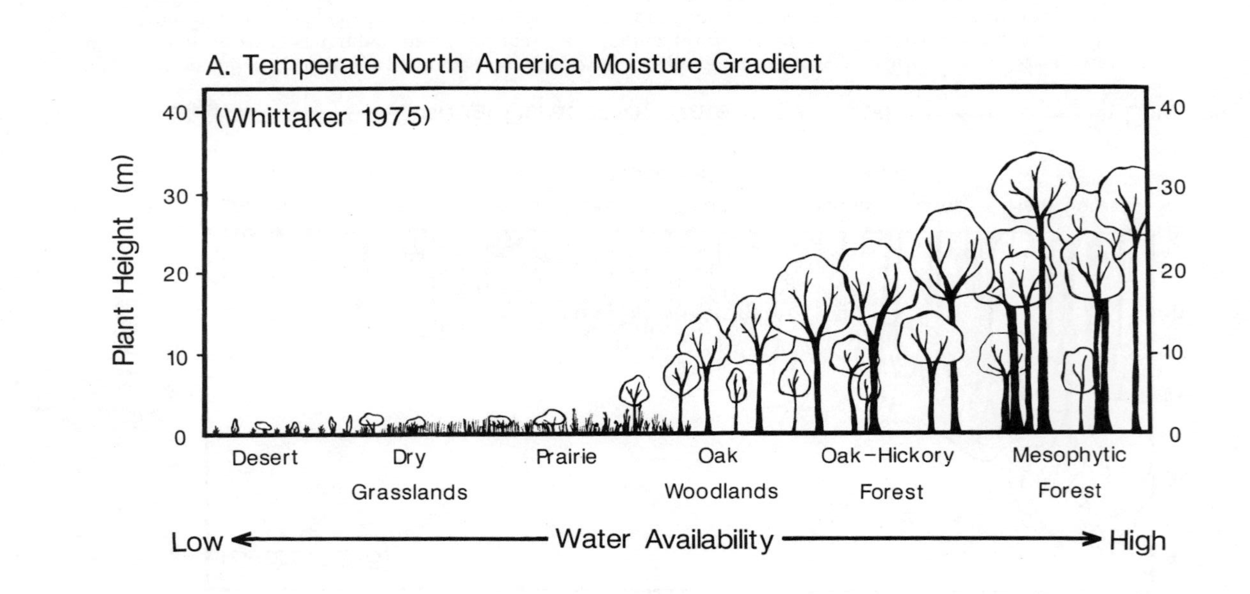
A. Temperate North America Moisture Gradient
(Whittaker 1975)
Plant Height (m)
40
30
20
10
0
Desert
Dry Grasslands
Prairie
Oak Woodlands
Oak-Hickory Forest
Mesophytic Forest
Low
Water Availability
High

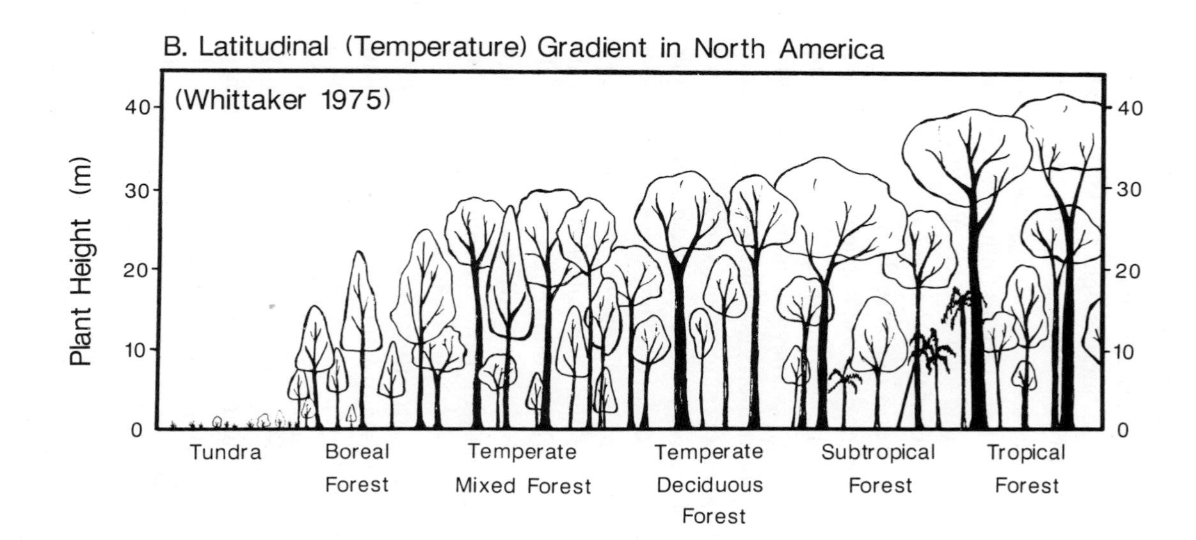

FIGURE 5.4. (A) Plant morphologies along a gradient from desert to mesophytic forest in temperate North America. This west to east gradient at mid-latitudes is mainly a gradient in rainfall and thus water availability, but also includes different fire frequencies. Modified from Whittaker (1975). (B) A latitudinal gradient in North America from habitats with low temperatures and short growing seasons to areas with high temperatures and long growing seasons. Modified from Whittaker (1975).

ronmental factors that reduce productivity or standing crop also favor species of shorter stature. For few of these gradients is there much experimental evidence as to what the actual limiting soil resources might be. Further, all of these are complex environmental gradients along which numerous physical characteristics change. Some, but not all, of these physical characteristics influence the supply rates of potentially limiting soil resources. For instance, the montane series has cooler temperatures at higher elevations. Cooler temperatures decrease the rate of microbial decomposition, and thus decrease nutrient mineralization rates. Similarly, decomposition slows as soils become increasingly anaerobic, as they do along the seasonal-swamp and swamp formation gradients. Low soil moisture also slows microbial decomposition and thus nutrient cycling and supply rates. Thus, all of these gradients may have a component in common: they are gradients along which the rate of supply of one or more limiting soil resources may increase and the availability of light (at the soil surfaces and at various distances below the top of the canopy) decrease. They all thus have a soil-resource:light gradient embedded in them. However, some of the factors that change along the gradients, such as temperature, also directly influence plant growth rates, independent of their effects on the supply rate of limiting soil resources. Thus these broad scale environmental gradients cannot be unambiguously interpreted as soil-resource:light gradients, even though that may be a major factor leading to their structure. In total, the qualitative physiognomic characteristics of the vegetation along these gradients is consistent with the predictions of Chapter 4, as is the continuous intergradation of species along each of these gradients.

These patterns of species replacements and plant heights along soil-resource:light gradients have a consistent element worldwide. These broad, consistent patterns that occur over a wide range of spatial scales worldwide indicate the existence of a major environmental constraint that has shaped plant evolution and the structure of terrestrial ecosystems. For vascular plants, these patterns are consistent with plants being specialized on different regions along a soil resource: light gradient, with plant morphology being a major determinant of competitive ability along these gradients.

NUTRIENT SUPPLY RATES

I know of only one case in which actual rates of supply of a soil nutrient have been measured *in situ* and related to the productivity and composition of edaphic climax vegetation, and that is the work by Pastor, Aber, McClaugherty, and Melillo (1984) on Blackhawk Island. Blackhawk Island, a 70 hectare island in the Wisconsin River, contains a series of old-growth, never-logged forest stands that occur on a variety of soils ranging from sands to clay loams to organic soils. These different forest stands, which include stands dominated by red pine, white pine, red oak, white oak, maple, or hemlock, represent a series of edaphic climaxes. Nitrogen is the most likely limiting soil nutrient in all of these stands (Pastor et al. 1984).

In each of eight different stands, Pastor et al. (1984) determined the total seasonal rate of nitrogen mineralization (the rate at which the mineral forms of nitrogen, NH_4 and NO_3, are made available in a soil), using an *in situ*, buried bag technique, as well as the seasonal net rate of above-ground primary productivity. They found that primary production was significantly correlated with annual nitrogen mineralization (Fig. 5.5A) and also with the percent of silt and clay in the subsoils from each of the forest stands (Fig. 5.5B). The silt and clay content of the subsoils is indicative of the parent material on which each of these soils formed. Thus, parent material with high silt and clay content led to soils with high rates of nitrogen mineralization

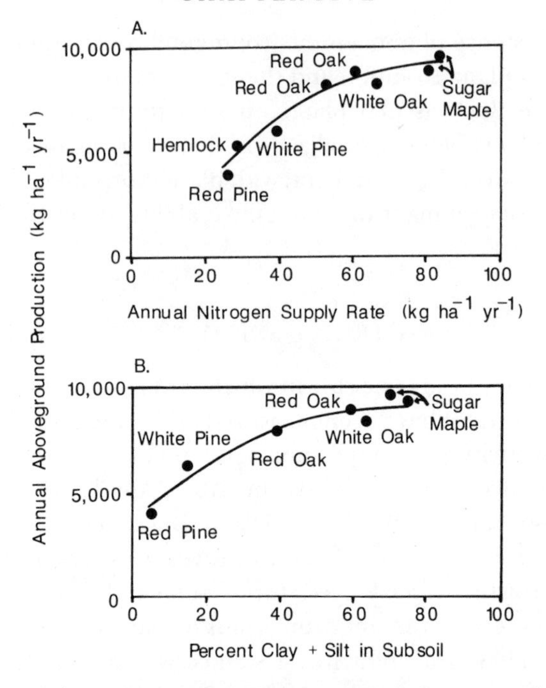

FIGURE 5.5. (A) Relationship between annual above-ground productivity of 8 forest stands on Blackhawk Island, Wisconsin, and the measured annual *in situ* rate of nitrogen mineralization. (B) Dependence of productivity of 7 of these stands on the clay and silt content of the subsoil. The subsoil should be indicative of the original parent material on which the soil formed. The hemlock stand was omitted because it lacked a subsoil. Redrawn from Pastor et al. (1984).

and high rates of primary productivity, whereas sandy parent material led to soils with low rates of nitrogen mineralization and low productivity.

Let's consider these vegetation stands in order from those occurring on the sandiest subsoils (least subsoil silt and clay) to those on the least sandy subsoils. Note that hemlock is not included in Figure 5.5B because it lacked a mineral subsoil, having 59% organic matter in its soil (Pastor et al. 1984). Red pine (*Pinus resinosa*), which foresters (Fowells 1965) rate as "shade intolerant" (i.e., as a poor competitor for light), dominated the areas with the sandiest soils and the lowest rate of nitrogen mineralization. Although I do not know its heights at Blackhawk Island, it tends to be 16–22 m tall at maturity in the region. The next least sandy soil, which had a higher rate of nitrogen mineralization, was dominated by white pine (*Pinus strobus*). White pine is considered to be of intermediate shade tolerance, i.e., as being of intermediate competitive ability for light. White pine can grow to great heights, but most are 18–24 m tall in the region. Next came two stands dominated by red oaks (*Quercus rubra*), which occurred on sandy clay loams that had even higher rates of nitrogen mineralization. Red oaks are rated intermediate in their shade tolerance (Fowells 1965). The white-oak (*Quercus alba*)-dominated stand had a subsoil with higher silt and clay and a higher rate of nitrogen mineralization than the red oak stands. White oak is rated shade tolerant, i.e., it is a good competitor for light. The two stands dominated by sugar maple (*Acer saccharum*) had the highest rates of annual nitrogen mineralization and the most silt and clay in their subsoils. Sugar maple is rated as the most shade tolerant species in the region, and thus should be the best competitor for light. It attains heights of from 22 to 30 m. Pastor et al. (1984) did not report light penetration to the soil surface for these stands. However, John Pastor has kindly provided me with data on light intensities at the soil surface that he collected in each of these stands (Fig. 5.6). Red pine, which occurs on the soil with the lowest rate of nitrogen mineralization, has the greatest penetration of light to the soil surface. Sugar maple, which occurs on the soils with the highest rates of nitrogen mineralization, has the lowest penetration of light to the soil surface. Indeed, there is a highly significant negative correlation between the annual rate of nitrogen mineralization in these seven stands and the intensity of light at the soil surface on Blackhawk Island. This negative correlation occurs using either the rate of nitrogen miner-

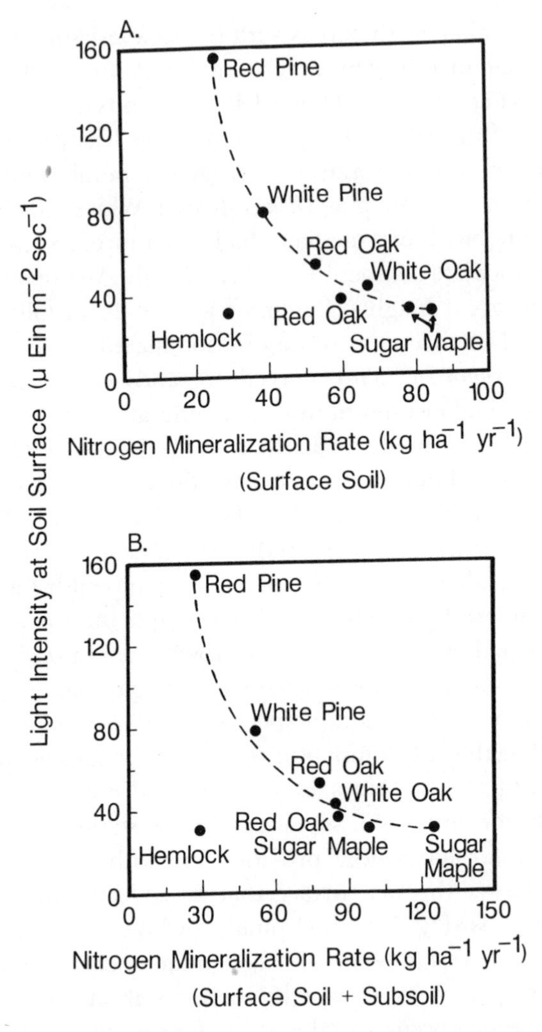

FIGURE 5.6. (A) Relation between measured light intensity at the soil surface and annual measured nitrogen mineralization rate for the 8 forest stands on Blackhawk Island, Wisconsin, using annual nitrogen mineralization rates estimated from just surface soils. (B) An essentially identical relationship using annual mineralization rates based on both surface and subsurface soils. I thank John Pastor for allowing me to use his unpublished light data.

alization in the surface soils (Fig. 5.6A) or the total rate of mineralization for both surface and subsurface soils (Fig. 5.6B). Thus these seven forest stands fall at different points along a soil-resource:light gradient. Their position along this gradient, which depends on the initial parent material upon which the soils formed, corresponds well with the ranking of their competitive ability for light, as given by the shade tolerance ratings of forestry. These results are consistent with the hypothesis that there is a tradeoff between the nitrogen competitive abilities and the light competitive abilities of these dominant tree species. It is inconsistent with Grime's (1979) concept of a "unified competitive ability," which asserts that a species that is a superior competitor for light is also a superior competitor for nitrogen, phosphorus, etc.

The one stand not consistent with this pattern is that dominated by hemlock. Eastern hemlock (*Tsuga canadensis*) is a highly shade tolerant species (Fowells 1965). However, the soils on which it occurred had very low rates of nitrogen mineralization. Pastor et al. (1984) omitted it from their version of Figure 5.5B because the stand lacked a subsoil and had a highly organic soil. Hemlock is unusual in several other ways, including the high tannin content of its bark and needles (Fowells 1965). Indeed, its bark is so high in tannin that it was selectively logged and stripped of its bark for commercial tannic acid production. It may be that these high tannin levels inhibit nitrogen mineralization in these soils. Its shade tolerance suggests that hemlock may be able to invade areas with nutrient-rich soils, but its occurrence on soils with low rates of nitrogen mineralization suggests that adults, at least, can survive on poor soils. It may be that this seeming contradiction is caused by the tannin content of the litter produced by hemlock. Hemlock may invade an area that has suitable sites for seedling establishment, and grow into the dominant species. However, the high tannin content of its litter may cause it eventually to have beneath it a soil with sufficiently low rates of nitrogen mineralization that its own regeneration is inhibited. Such an effect could explain its frequent failure to regenerate after logging, and could cause long-term hemlock cycles on a given site.

The hemlock example makes a point that is not always appreciated by population ecologists, but which is almost the creed of ecosystem ecologists: no species or process exists in isolation from other species or processes in an ecosystem; rather, all have the potential to influence and interact with each other. Such feedback can tend to stabilize or to destabilize a given interaction. For seven of the eight stands that Pastor et al. (1984) studied, the long-term feedback of litter production on soil mineralization rates seems to stabilize the interactions, leading to a persistent, stable, vegetational composition that is closely related to the original parent material on which each soil formed. Thus these stands are considered to be "edaphic," or soil-caused, climaxes. This dependence is clearly illustrated by the correlation between the subsurface soils' silt and clay content and the ultimate composition of the dominant vegetation. Very sandy soils, with little subsurface silt and clay, drain easily and lack cation exchange sites that can bind NH_4. Thus, such parent materials are likely to have high leaching loss rates of nitrogen, and ultimately equilibrate at a rather low nitrogen level (Olson 1958). Parent material with high clay and silt content will have lower leaching losses of nitrogen, and thus ultimately develop soils with higher nitrogen availabilities, just as happened on Blackhawk Island. However, the pattern is also influenced by the plants living on a site, and the "quality" of their litter. Plants producing readily decomposed litter with a low carbon:nitrogen ratio will support a more luxuriant decomposer community, and thus have a more rapid rate of nitrogen mineralization. If such plants tend to be species that are dominant on rich soils, this could lead to a stabilizing feedback loop in which plants produce litter that produces soils that favor their own existence. Just this seems to be happening with sugar maple, which produces high-nitrogen leaves that are rapidly decomposed (Pastor et al. 1984). If plants dominant on poor soils produce litter that decomposes slowly and has a high carbon:nitrogen ratio, mineralization rates would be low, and such species would tend to maintain environmental conditions that were favorable to their continued existence. Red pine does just this (Pastor et al. 1984). Although much more work needs to be done on its biology, hemlock may require relatively rich soils for its establishment, and yet it eventually generates a poor, acidic soil with low rates of mineralization because of the acids and refractory organic compounds in its litter.

Are such patterns mere happenstance, or might the litter quality of a plant be an evolved trait, which should thus be frequently correlated with a plant's nutrient competitive ability? Although there is much that we need to understand before we can answer these questions, it seems possible for selective differences among individuals to lead to changes in the quality of the litter they produce, and for plants to have their competitive abilities evolve to be consistent with the soils their litter indirectly helps produce. However, such an evolutionary argument hinges on the viscosity of the environment, i.e., on each individual mainly influencing its immediate environment and the environment experienced by its offspring. Thus, most of the litter produced by a particular individual plant must fall beneath that individual, and affect the soil in which it and its offspring are rooted. Considering the shape of "seed shadows" (Janzen 1970), this seems plausible. This would then suggest that a species that forms a "climax" stand should produce a litter that reinforces the soil nutrient supply rates for which it is a superior competitor. However, the close correspondence between original parent material and the ultimate "edaphic climax" vegetation of an area demonstrates that such litter feedback effects are not the main determinant of the ultimate vegetational composition. Sugar maple seeds have surely invaded the pine and oak stands of Blackhawk Island many times during the last thousand years, and yet sugar maple, despite the high nutrient availability of the litter it produces, has not displaced the pine or oak. Alternatively, though, a species that produces a litter that is inconsistent with its competitive abilities, such as hemlock may do, may be unable to maintain a long-term, persistent stand on an area.

There are alternative explanations for the correlation between litter quality and the nitrogen supply rate of the soil on which a species is dominant. For instance, for plants dominant on poor soils, those individuals that minimize their loss of nitrogen will be favored over those that do not (Chapin 1980). A simple mechanism that minimizes nitrogen loss is translocation of nitrogen from a leaf before the leaf is shed. The more important nitrogen is to a plant, the greater would be the selective advantage it received from such translocation. Species dominant in areas with nitrogen-poor soils would be likely to produce litter that had a higher carbon:nitrogen ratio than species dominant in more nitrogen-rich habitats, assuming that translocation has a cost. It may be that the effect of different litter carbon:nitrogen ratios on soil nitrogen supply rates are an evolutionarily indirect effect of selection for optimal nitrogen conservation in individual plants. Whatever the evolutionary explanation, the work on vegetation on Blackhawk Island has demonstrated a clear link between original parent material and vegetation, and shown that the relations between soil nutrient supply rates and light intensity at the soil surface that were hypothesized in Chapter 4 do actually exist in natural vegetation.

Shade Tolerance: The forestry concept of "shade tolerance," which I have mentioned above, is a measure of the ability of seedlings and saplings to survive and grow at low light levels. It is in this stage in their life cycle that trees experience their lowest levels of light availability, and thus "shade tolerance" may be a measure of competitive ability for light. Unfortunately, the term "tolerance" does not include any implication of intra- or interspecific interaction, but rather suggests that species differ in their ability to survive some sort of environmental harshness. However, the different shade levels that trees must "tolerate" are actually different levels of availability of a potentially limiting resource, light, that are determined by the biomass and height distributions of other plants. Light attenuation, despite its vertical component, is not qualitatively different from any other act of resource consumption. As such, it should be realized that "shade tolerance" is just one component of the process of light competition. An understanding of plant competition requires that we go beyond the concept of tolerance by studying how physiological and morphological traits of plants of various size classes determine the competitive interactions of species.

SEED SIZE

Seed size is another trait of a plant that the theory of Chapter 4 predicted should be influenced by the position of a plant along soil-resource:light and loss rate gradients. Larger seed size was favored in nutrient-rich habitats with low loss rates because larger seeds led to taller seedlings, and taller seedlings received more light. Small seed size was favored in nutrient-poor habitats or habitats with high loss rates. Many other factors not included in the model can also influence seed size. For instance, if seed or seedling predators were to keep seedling density sufficiently low that there was little vertical light gradient near the soil, then larger seeds would not be favored on richer soils.

Seasonality could favor larger seeds if early, rapid growth could allow an individual to obtain more of a resource available for only a short time, such as water in a desert, or to pre-empt a resource such as light by growing taller, faster, than other individuals. A thick litter layer could also favor larger seeds, since larger seeds would have the necessary reserves to produce roots that could penetrate into the soil and leaves that could grow above the litter into the light.

Grime and Jeffery (1964) found, for nine North American tree species, that the maximum height attained after 12 weeks of seedling growth in dense shade was highly positively correlated with seed weight. Thus, larger seeds did lead to taller seedlings, as assumed by ALLOCATE. In an intraspecific competition experiment performed among seedlings of sunflowers (*Helianthus annuus*), seedlings from large seeds increased in height more quickly and thus outcompeted seedlings from small seeds when grown in a dense stand (Kuriowa 1960). Salisbury (1942) reported seed sizes for 377 species of British plants. He classified these species as to where each occurred along a gradient from open, high light habitats to closed canopy, low light habitats. The seven categories he used were (1) species of open habitats; (2) species of semi-closed (non-shady) habitats; (3) meadow species; (4) herbs of scrub and woodland margins; (5) ground flora of woodlands; (6) shrubs; and (7) trees. His results, summarized in Figure 5.7A, show that average seed size increased with increasing light limitation. He associated closed canopy habitats with late succession, and stated that "the larger supply of food material provided by the parent plant in the seed, or other propagule, the more advanced the phase of succession that the species can normally occupy" (Salisbury 1942, p. 18). Baker (1972) ana-

FIGURE 5.7. (A) Relationship between seed size of various plant species and the approximate point on a soil-resource:light gradient at which each species is most abundant, using 287 British plant species. Mean and standard errors of seed sizes are based on the data in Salisbury (1942), using the species classification as to habitat openness (light availability) given by Salisbury. (B) Similar data for 2160 species of California plants, using data from Baker (1972), with species classified as to life form by Baker. Note that seed size is logarithmic.

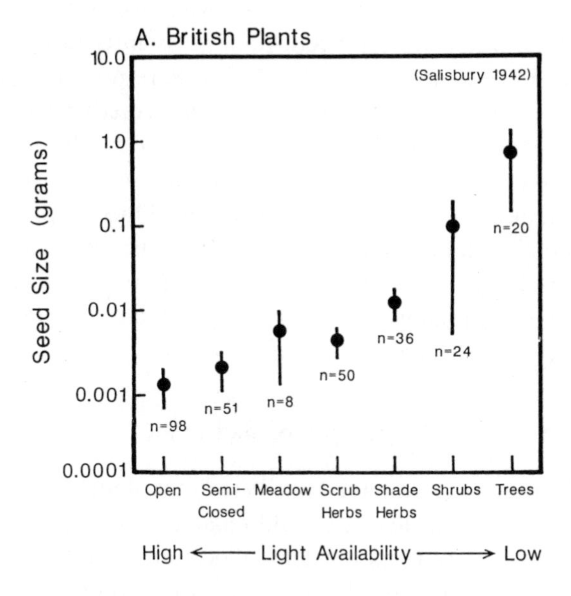
A. British Plants
(Salisbury 1942)
Seed Size (grams)
10.0
1.0
0.1
0.01
0.001
0.0001
n=98
n=51
n=8
n=50
n=36
n=24
n=20
Open
Semi-Closed
Meadow
Scrub Herbs
Shade Herbs
Shrubs
Trees
High ← Light Availability → Low

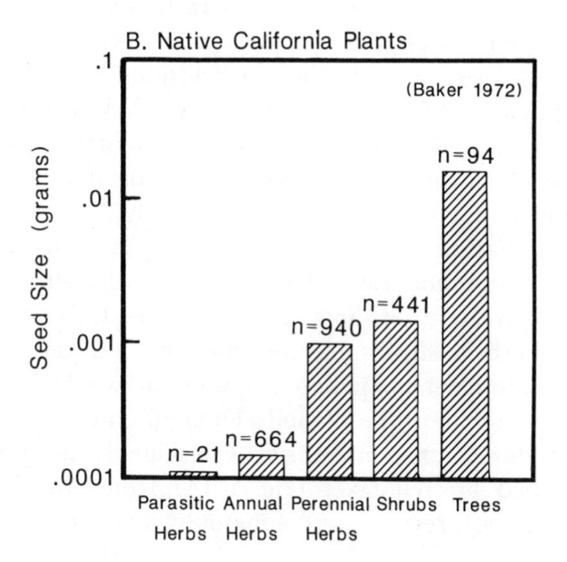
B. Native California Plants
(Baker 1972)
Seed Size (grams)
.1
.01
.001
.0001
n=21
n=664
n=940
n=441
n=94
Parasitic Herbs
Annual Herbs
Perennial Herbs
Shrubs
Trees

lyzed seed sizes of 2160 species of native Californian plants. Trees, which dominate habitats with richer soils and a more mesic climate, have large seeds, whereas annual herbs, which dominate dry, nutrient-poor grasslands, deserts, and scrub, have small seeds (Fig. 5.7B). Perennial herbs and shrubs, which are most common in intermediate habitats, have intermediate seed sizes. Baker (1972) showed that other environmental variables, including moisture and elevation, also had an impact on seed size. Thus, as predicted by theory, plants characteristic of habitats with poor soils and high light penetration have much smaller seeds, on average, than plants characteristic of habitats with rich soils and low light availability at the soil surface.

ALLOCATION PATTERNS

The theory of Chapter 4 predicted that plant allocation to roots, stems, and leaves should change along productivity gradients, with maximal leaf allocation at intermediate points on such gradients, and with root allocation decreasing but stem allocation increasing along gradients from poor to rich soils. A variety of observations are generally consistent with these predictions. For instance, Chapin (1980) found that plant species dominant in areas with poor soils invest more heavily in mycorrhizal associations (which function as roots) and in roots than do plants dominant in areas with rich soils. Similarly, Mooney (1972) stated that annuals and perennial herbs tend to have greater root:shoot ratios than shrubs, and that shrubs tend to have greater root:shoot ratios than trees. If the sequence from herbs to shrubs to trees represents a sequence from species dominant on poor to rich soils, this would support a greater allocation to roots and a lower allocation to stems in plants dominant on poorer soils. Consider also the gradient described by Whittaker (Fig. 5.4A) from Appalachian mesophytic forest to desert. A mesophytic cove forest (dominated by tulip trees, *Liriodendron tulipifera*), with a net primary productivity of 1300 g/m^2/yr, had a root:shoot ratio of 0.16:1 (Whittaker 1975; his Table 5.1). A less productive oak-pine forest (1060 g/m^2/yr) had a root:shoot ratio of 0.52:1 (Whittaker 1975). Temperate grassland (prairie), with an average productivity of 600 g/m^2/yr (Whittaker 1975, his Table 5.2), had a root:shoot ratio of about 6.5:1 (Barbour, Burk, and Pitts 1980). (This high ratio may partially result from periodic fire and mammalian grazing, both of which would favor increased below-ground allocation.) Although I have not found exact root:shoot ratios for desert plants, Cody's (1986) interesting discussion of patterns of above- and below-ground life forms in desert plants suggests that many long-lived desert plants may have root:shoot ratios greater than 2. Whittaker reports that desert has a mean productivity of about 90 g/m^2/yr. Thus, the gradient of Figure 5.4A is a productivity gradient along which there is higher allocation to roots on resource-poor soils and higher allocation to stems on resource-rich soils.

Within species, individual plants have higher root:shoot ratios in habitats in which they are limited by water or a mineral nutrient, and lower ratios in high nutrient and/or low light habitats (Mooney 1972; Grime and Hunt 1975; Shamsi and Whitehead 1977; Grime 1979; Parrish and Bazzaz 1982; Bloom, Chapin and Mooney 1985; Chapin and Shaver 1985). Thus, both among species and within species, allocation to roots versus allocation to shoots tends to vary along a soil-resource:light gradient in a manner consistent with expectations of optimal foraging theory for essential resources (Rapport and Turner 1975; Tilman 1982; Bloom, Chapin, and Mooney 1985; Chapter 2 this volume).

However, root:shoot ratios are only an approximate index of allocation patterns for the acquisition of below-ground versus above-ground resources. Roots function not only as structures for the acquisition of soil resources, but also as support structures and as storage organs. In plants such as perennial grasses and forbs in which much of the shoot is reconstructed each year, the resources used for this reconstruction are stored in the stem bases and roots (Mooney 1972). Such species must allocate to storage compounds in order to grow during the next season. The high root:shoot ratios seen in many prairie species may be as much a result of storage as of increased effort being expended for nutrient uptake. Further, standing crop is not indicative of turnover or total investment in a function. Because roots may turnover much more rapidly than stems, and because plants may invest considerable production in mycorrhizal associations, root biomass is a poor estimate of allocation for the acquisition of below-ground resources. In addition, shoots contain both structural tissues, used to support leaves, and photosynthetic tissues. Tall plants require more support than shorter plants (Greenhill 1881, McMahon 1973), and must also allocate more to the production of below-ground structural tissues than shorter plants. For these reasons, root:shoot ratios are but a first approximation to allocation patterns to nutrient acquisition versus light acquisition. Despite these problems, the differences that have been reported are so great and the patterns so consistent that they must be considered strong support for the role of allocation tradeoffs in structuring plant communities.

PLANT ABUNDANCE AND MORPHOLOGY ON EXPERIMENTAL GRADIENTS

The gradients summarized in Figures 5.2–5.6 demonstrate major changes in vegetational composition and species morphology along a variety of complex environmental gradients that span geographically large distances. Although such broad, correlational patterns are striking in their similarity, and thus suggest that similar mechanisms may be at work in widely different habitats, all correlational patterns are open to alternative interpretations. To test the hypothesis that such patterns are caused by competition for light and soil resources, it is necessary to establish experimental gradients. If the responses to experimental gradients are similar to the patterns observed on natural, but more complex, environmental gradients, then the theoretical predictions would be supported.

There are several methods that could be used to establish such experimental gradients. One approach would be to determine experimentally what the limiting soil resource or resources were for a given habitat, and then to add this or these resources at a variety of different rates in replicated plots. If the theory of Chapter 4 is a suitable simplification of nature, higher rates of nutrient addition should lead to higher productivity, higher soil resource levels, decreased light intensity at the soil surface, and changes in allocation patterns and life histories. A second experimental approach would be to shade or artificially increase light levels. A third approach would be to manipulate both light and nutrients directly, perhaps in a full-factorial experiment, for natural vegetation. Adding nutrients should, in theory, lead to similar changes in plant morphology and vegetation composition as would reducing incident light levels, and vice versa. The third approach would provide the most experimentally direct test of the theory developed in Chapter 4. Unfortunately, there have not yet been any well-replicated studies that have proceeded long enough to allow the ultimate effect of an experimental soil-resource:light gradient, however it was experimentally created, to be determined.

The Park Grass Experiments of the Rothamsted Experimental Station, in Harpendon, England, have no replicates of most treatments, but have been in existence over 130 years, having been started in 1856 (Lawes and Gilbert 1880; Thurston 1969; see description in Tilman 1982). The 20 different experimental plots, each of which has received a particular combination of nutrients for the past 130 years, have shown dramatic changes in species composition following various patterns of nutrient addition and various manipulations of soil pH. These changes are consistent with the hypothesis that the major limiting resources are light and soil nutrients, especially nitrogen and phosphorus (Tilman 1982). However, because the plots are mowed for hay twice each year, the usual succession to forest which would occur on these soils has been prevented. This loss is not completely comparable to the loss analyzed in Chapter 4 because that was density-independent mortality that fell equally on all species. Mowing removes above-ground biomass, but does not necessarily kill plants. It selects much more heavily against woody and other tall growth form species than against herbaceous and short growth form species. Despite this, the qualitative predictions of expected changes in life histories along a soil-resource:light gradient (Fig. 4.14) should hold. Considering only those plots with soil pH greater than 4.5, the plots receiving complete nutrient addition with the highest rate of nitrogen supply were dominated by the tallest species, *Alopecurus* and *Arrhenatherum*. These are both stout, erect perennial grasses that have heights averaging about 60–80 cm. In contrast, the unfertilized control plots are dominated by shorter, tufted, branching, creeping or rosette-forming species such as *Festuca rubra*, *Agrostis tenuis*, and *Leontodon hispidus*, which have average heights of 15–30 cm.

The Park Grass Plots have thus shown that the addition of limiting soil nutrients, which increase total plant biomass and decrease light penetration to the soil surface, favors species that have a more erect growth form and which can attain greater heights. Snaydon (1970), Snaydon and Davies (1972), and Snaydon (1976) have shown that these forces have also favored a similar pattern of heritable variation within individual species. Populations of *Anthoxanthum* collected from heavily fertilized plots had evolved both greater height and greater yield of above-ground biomass per individual than plants collected from unfertilized plots or plots with low rates of nutrient addition (Fig. 5.8). Thus, just as an increased supply of a limiting soil nutrient led to dominance by taller species, so it favored genetically taller individuals within a particular species. These individuals are both taller and have greater above-ground biomass per individual. Although below-ground biomass was not reported, this suggests that 120 years of living in a habitat with high nutrient levels and high standing crop had favored individuals that allocated more of their production to leaves and stems and less to roots compared to the individuals that were living in the plots before fertilization. As an additional test of the role of evolution with respect to local soil conditions in these populations, Snaydon (1976) reported the results of transplant experiments. Plants that were collected in six different Park Grass plots were grown in a common garden in Reading for several years, and then transplanted back both to their original plots and to other plots. Plants transplanted back to their original plots "survived twice as long as those transplanted into contrasting 'alien' plots" and "the surviving plants also produced 50% more dry matter in their native plots. This confirms that plants are better suited to the conditions of their native plots. It also suggests that natural selection eliminates the less well suited plants and favors the best suited plants on each plot" (Snaydon 1976, pp. 16-17).

These results suggest that the major environmental variables that constrain and thus structure plant communities do so on four different levels. At the level of the individual plant, these gradients will favor phenotypic plasticity that will allow an individual to adjust its resource requirements and thus compete more effectively in a variety of different habitats. At the level of the population, these gradients will favor different genotypes in different habitats. At the level

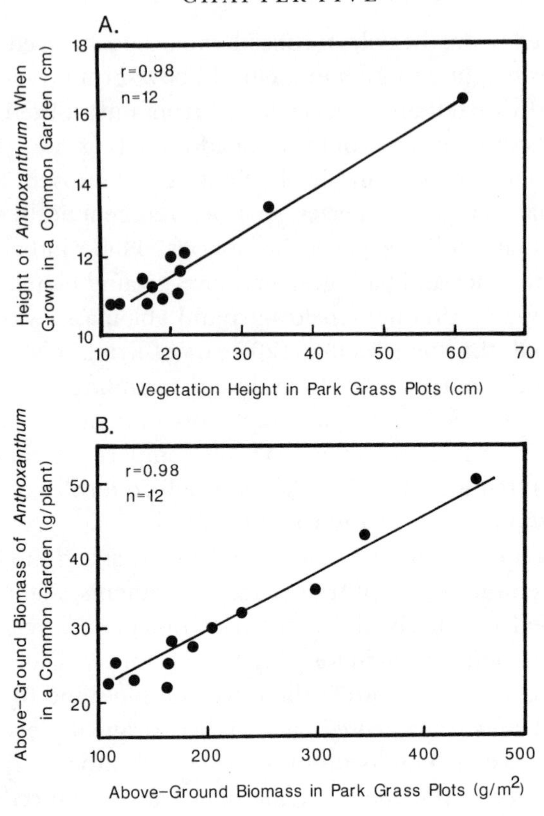

FIGURE 5.8. (A) Ecotypic (genotypic) variation in heights of *Anthoxanthum odoratum*, and its relation to vegetation heights in the Park Grass plots from which each genotype was obtained. Note that tall genotypes came from plots with tall vegetation. (B) Similar ecotypic variation in the above-ground biomass of this species in relation to the total above-ground biomass of Park Grass plots from which individual plants of *Anthoxanthum odoratum* were obtained. Redrawn from Snaydon (1976).

of the community, these gradients will favor some species over others in different habitats. At the level of biomes, these forces lead to grasslands, shrublands, savannas, and closed-canopy forests. Clearly, not all environmental factors will act similarly on all these levels of organization. However, it is those variables that do act similarly on all these levels that are likely to cause much of the similarity we see from habitat to habitat around the world.

In 1982 a series of replicated experimental nitrogen gradients was established in four different successional fields at Cedar Creek Natural History, Minnesota. By 1986, most of the common species were significantly separated along these experimental gradients, with the low nitrogen end of the gradients dominated by short, often rosette-forming herbs, whereas the high nitrogen end of the gradients was dominated by tall, erect species, including some woody species (Tilman 1987a). These experiments are discussed in Chapter 8.

CORRELATIONS AMONG TRAITS

I know of no habitat in which all the relevant morphological, physiological and life history traits have been measured on the dominant species, and in which these traits could be related to the position along a soil-resource:light gradient or a loss rate gradient at which each species reached its peak abundance. However, the work of Salisbury (1942) on seed sizes of plants of different successional stages can be combined with that of Grime and Hunt (1975) on maximal relative growth rates and with the life histories, plant heights, and habitat affinities reported in Clapham, Tutin, and Warburg (1962). These analyses showed that compared to perennials, annuals have higher *RGR*'s (1.7 versus 1.1 wk^{-1}), smaller seed size (0.0016 vs. 0.93 g/seed), and are shorter at maturity (38.5 vs. 339 cm). Woody perennials have the largest seed size (0.32 g/seed), the greatest height (1200 cm), and the slowest growth rate (0.60 wk^{-1}), on average, compared to all other groups of annuals and perennials. These patterns are consistent with the predictions of Chapter 4. In addition, Salisbury's categorization of habitat types for these species, in combination with those in Clapham, Tutin, and Warburg (1962), allows these data to be used to determine how plant height at maturity and maximal growth rate depend on the habitats in which species naturally occur (Fig. 5.9). Assuming that Salisbury's classification is truly indicative of succession, as he asserted, or of light availability at the soil surface, then plant heights and seed sizes increase as light availability decreases, and maximal growth rates decrease. I don't know the extent to which the gradient of Figure 5.9 represents a loss rate gradient versus a productivity gradient. I am inclined to interpret it as a loss rate gradient, going from highly disturbed, "early successional" habitats to less disturbed, "late successional" habitats. I base this on the points raised in my earlier discussion of Tansley (1949), Godwin (1956), and Grime (1979). If my interpretation is correct, Figure 5.9B shows that, as predicted (Fig. 5.1), RGR_{max} increases with loss rate. Lumping all categories together, there are some other interesting patterns. Seed weight and plant height at maturity are highly significantly correlated (Fig. 5.10A). Maximal relative growth rate is negatively correlated with seed weight (Fig. 5.10B). RGR_{max} is negatively correlated with height at maturity, but not significantly so (Fig. 5.10C). Although there is much variability in these data, these correlations suggest that plants that are taller at maturity tend to have larger seeds and slower maximal growth rates than plants that are shorter at maturity, just as predicted in Chapter 4.

PHENOTYPIC PLASTICITY

The height, shape, age at first reproduction, branching pattern, and other morphological features of plants are

FIGURE 5.9. (A) Dependence of plant height at maturity, based on data in Clapham, Tutin, and Warburg (1962), on habitat type in which each species is most abundant. (B) Maximum relative growth rate, based on data in Grime and Hunt (1975), as dependent on habitat type. (C) Dependence of seed size, based on data in Salisbury (1942), on the type of habitat in which each species is most abundant. Habitat type is based on Salisbury (1942) and Clapham, Tutin, and Warburg (1962). All relations are for plants of the United Kingdom. Mean and standard error are shown for each case.

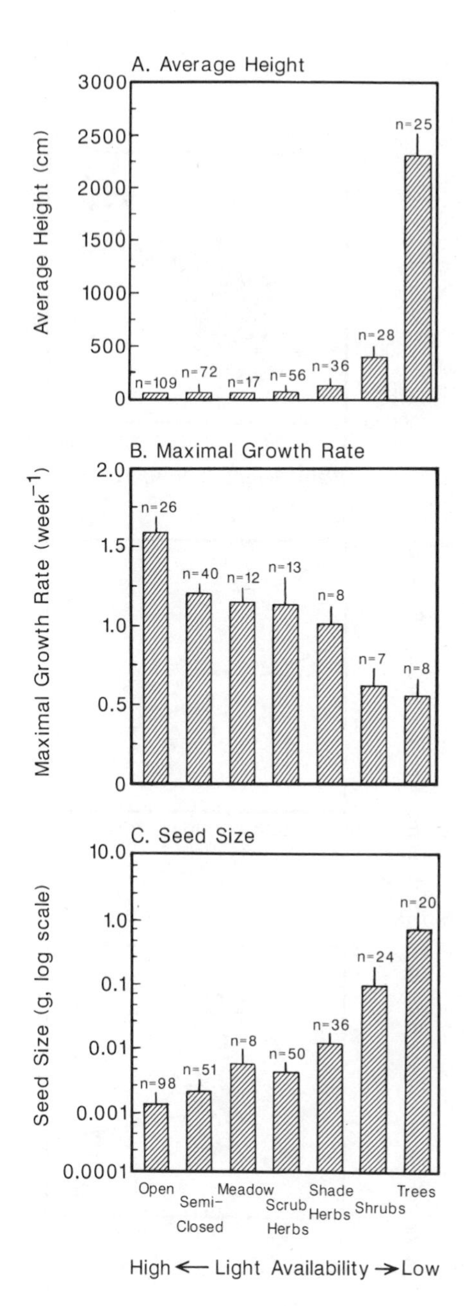

A. Average Height
Average Height (cm)
3000
2500
2000
1500
1000
500
0
n=109
n=72
n=17
n=56
n=36
n=28
n=25
B. Maximal Growth Rate
Maximal Growth Rate (week^{-1})
2.0
1.5
1.0
0.5
0
n=26
n=40
n=12
n=13
n=8
n=7
n=8
C. Seed Size
Seed Size (g, log scale)
10.0
1.0
0.1
0.01
0.001
0.0001
n=98
n=51
n=8
n=50
n=36
n=24
n=20
Open
Semi-Closed
Meadow
Scrub Herbs
Shade Herbs
Shrubs
Trees
High ← Light Availability → Low

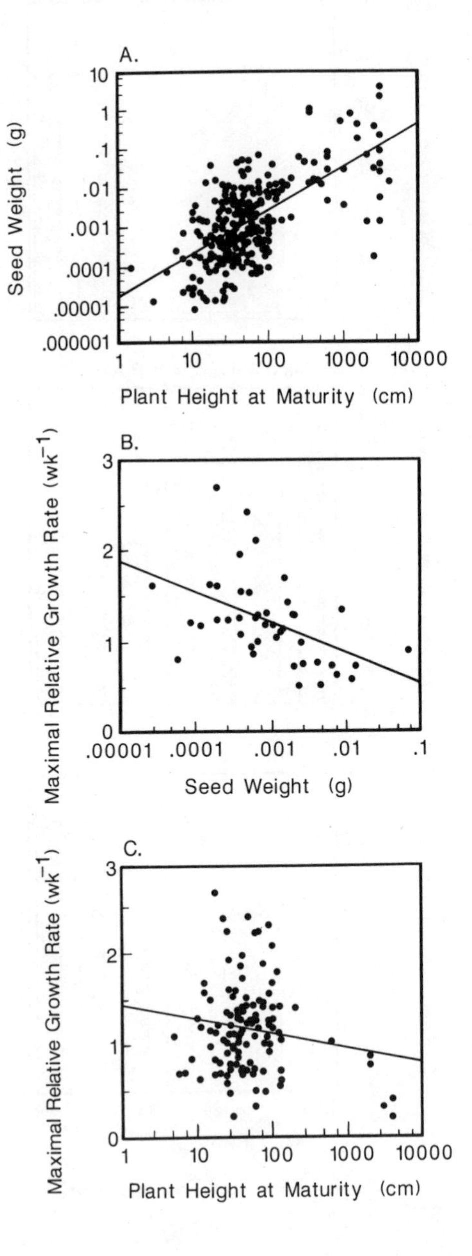
A.
Seed Weight (g)
10
1
.1
.01
.001
.0001
.00001
.000001
1
10
100
1000
10000
Plant Height at Maturity (cm)
B.
Maximal Relative Growth Rate (wk^{-1})
3
2
1
0
.00001
.0001
.001
.01
.1
Seed Weight (g)
C.
Maximal Relative Growth Rate (wk^{-1})
3
2
1
0
1
10
100
1000
10000
Plant Height at Maturity (cm)

highly dependent on the environmental conditions under which they grow. As discussed by Harper (1977) these morphological differences come most from changes in the numbers of various plant units (leaves, flowers, etc.), and not from major morphological changes in the units themselves. When grown at low densities on a nutrient-rich soil, plants tend to be short and bushy, with a multi-layered leaf canopy. In contrast, when grown at high densities on a rich soil, plants are taller and have more of their leaves near the top of the plant (Horn 1971). One cause of this is that the older, lower leaves of plants growing at high-density do not receive enough light to have their rate of photosynthesis equal their respiration rate (Harper 1977), and are lost. Thus, plants allocate more of their growth to structural tissues associated with height when growing under high density conditions for which light attenuation is great. Additionally, such high-density, light-limited situations lead to "reduced reproductive output, reduced growth rate, delayed maturity and reproduction" (Harper 1977, p. xvi), just as was predicted in Chapter 4.

Although leaves are a relatively stable morphological unit in most plants compared to the morphological changes in the entire individual, the photosynthetic capabilities of leaves depend on their exposure to sun. Leaves that receive full sunlight, called sun leaves, tend to have higher light-saturated (maximal) rates of photosynthesis than leaves growing in the shade. Shade leaves, however, tend to have

---

FIGURE 5.10. (A) Log of mass per seed and log of plant height at maturity are significantly correlated ($r = 0.64$, $n = 274$, $P < 0.01$), using the data from Salisbury (1942) and Clapham, Tutin, and Warburg (1962). (B) Maximal relative growth rate is significantly negatively correlated with log of mass per seed ($r = -0.48$, $n = 43$, $P < 0.01$), using data from Grime and Hunt (1975) and Salisbury (1942). (C) Maximal relative growth rate is not significantly linearly dependent on log of height at maturity ($r = -0.16$, $n = 123$, $P > 0.05$), using data from Clapham, Tutin, and Warburg (1962) and Grime and Hunt (1975). The residuals to this regression suggest a somewhat quadratic relationship, with plants of intermediate height having the greatest maximal growth rate.

their rate of photosynthesis saturate at lower light intensities (Mooney 1972; Harper 1977; Lassoie et al. 1985).

All of these patterns of morphological and physiological plasticity within genetically identical plants are consistent with the hypothesis that this plasticity is an adaptation to competition for light and a limiting soil resource. There is an immense and interesting literature documenting the environmental physiology and morphology of plants (see papers in book edited by Chabot and Mooney 1985 and review in Bloom, Chapin, and Mooney 1985) that demonstrates the adaptive nature of individual plasticity and its frequent similarity to patterns of variation among species. One of the major problems with predicting the interactions of plants is that their great morphological variability makes it difficult to determine their competitive abilities in an unambiguous manner. Indeed, their competitive abilities are habitat dependent, as are their rates of acquisition of various limiting resources. However, whatever the cause, the range of variation within individuals is less than that among different members of a species, and the range of variation within a species is much less than the range of variation among species.

SUMMARY

In total, this review has shown that many aspects of the life history and morphology of plants depend on the position at which a species attains its peak abundance along a soil-resource:light gradient or along a loss rate gradient. For a wide variety of productivity gradients that have been described—be they gradients from habitats with low to high rates of nitrogen supply, or high to low elevations, or low to high latitudes, or low to high precipitations—there are strikingly similar changes in the morphology and life histories of the dominant plant species. The same types of changes that occur from one species to the next along such broad-scale environmental gradients occur, on a smaller scale, as genetic variation in these traits within individual species. Further, each individual plant is morphologically and physiologically plastic. This plasticity often has the same qualitative pattern of variation along such gradients as is seen for genotypic variation within and among species. Thus, for a wide range of plant traits—heights, allocation patterns to roots, leaves, and stems, seed sizes, growth rates, and photosynthetic efficiencies—it seems that soil resource supply rates and loss rates, which together determine the availabilities of soil resources and light, have been major axes along which there has been genotypic specialization within a species, speciation, and broad-scale convergence of phylogenetically distant plants living in climatically and edaphically similar habitats around the world. Such correspondence across this wide variety of scales suggests that the average availabilities of soil resources and light within a habitat, and the tradeoffs plants face in exploiting a soil resource versus light, have been major determinants of vegetational patterns.

CHAPTER SIX

The Dynamics of Plant Competition

Up to this point, I have mainly discussed the long-term outcome of plant competition, i.e., the eventual fate of various species once competitive interactions have reached an equilibrium. For instance, in discussing competition between two species under a particular set of environmental conditions, I have called a species a superior competitor if, at equilibrium, it competitively displaced the other species. I have emphasized the long-term, eventual outcome of competition (competitive dominance or coexistence) because it is this aspect of competition that is most likely to influence the broad patterns of geographic distribution and local abundance that we see in those natural plant communities that have escaped major human disturbances. Because such communities are our best reference point for the conditions in which current species evolved, because they harbor the riches of genetic diversity that has resulted from 3.5 billion years of evolution, and because they are fast disappearing, these natural, undisturbed communities are a treasured resource. However, the thinking of ecologists is highly flavored by the habitats they experience most frequently. Natural plant communities that are free from major human disturbances are increasingly rare on the earth's surface. The vast majority of the vegetation of the earth has been greatly modified by forestry, farming, grazing, and other anthropogenic disturbances. Furthermore, all ecological experiments are, of necessity, disturbances. Thus, much of what ecologists see are the highly dynamic, short-term responses of particular species following disturbance. The patterns seen over these short time scales are often extrapolated to draw inferences about the processes that may have been important in the evolutionary past of these species. As this chapter will show, such extrapolations are dangerous because the immediate response to an experimental manipulation can be transient dynamics that are unrelated to the long-term effect.

The purpose of this chapter is to explore some of the similarities and differences between the patterns we are likely to see in recently disturbed communities and those we are likely to see in natural communities that have been free of human disturbance. One of the themes developed in this chapter is that the immediate dynamic response to a particular disturbance or experimental manipulation can be markedly different from its long-term effect. Thus, the patterns and processes that seem most important in recently disturbed ecosystems may be of less importance in undisturbed ecosystems. The immediate dynamic responses of species to various disturbances or experimental manipulations may be evolutionarily indirect effects of other selective forces.

Within the framework of a particular model, it is a straightforward process to determine, either analytically or numerically, what the predicted outcome of competition should be. However, it is not immediately obvious what the best experimental or observational approaches might be to determine which species are actually the best competitors in a particular natural environment. With short-lived organisms, such as planktonic algae, it is easy to allow competition to proceed for 10 or 20 generations, and unambiguously determine the outcome of their interactions. However, many terrestrial plants have lifespans that are much longer than research grants, attention spans of scientists, or even lifespans of scientists. Almost all the work on competition among terrestrial plants has thus used various short-cuts that are designed to estimate the eventual, long-term outcome of competition using short-term experimental or observational data. These short-cuts, especially those based on the deWit replacement series or its derivatives, have given a different meaning to "competitive ability" than I have ascribed to it. These approaches use the short-term relative yields of two species (often just yields of aboveground biomass) grown at different relative densities as being synonymous with their "competitive abilities." Another approach that has often been taken in the literature has been to call a species a superior competitor if, in multispecies mixtures, it grows more rapidly than another species. Thus, if species *A*, *B*, and *C* were sown into a field, and species *A* was observed to increase in its abundance relative to *B* and *C* during the interval of observation, species *A* would often be called a superior competitor for the conditions that existed in that field. In contrast, I call a species a superior competitor for a particular habitat if it can eventually competitively displace other species from that habitat.

These terminological differences raise some important questions. Can the short-term dynamics of competitive interactions be used to predict their eventual long-term outcome? Should the species that have the highest initial relative yields eventually competitively displace the other species? Is there an expected pattern to the dynamics of competition?

SIMPLE MODELS OF COMPETITION

Let us first consider a case in which three idealized species which lack size or age structure compete for a single limiting resource (Fig. 6.1). When all species are experiencing the same loss rate, rate D_1, species *A* has the lowest R^*, and thus is the superior competitor for the resource. However, because species *A* has a lower maximal growth rate than species *B* and *C*, *B* and *C* will grow more rapidly initially

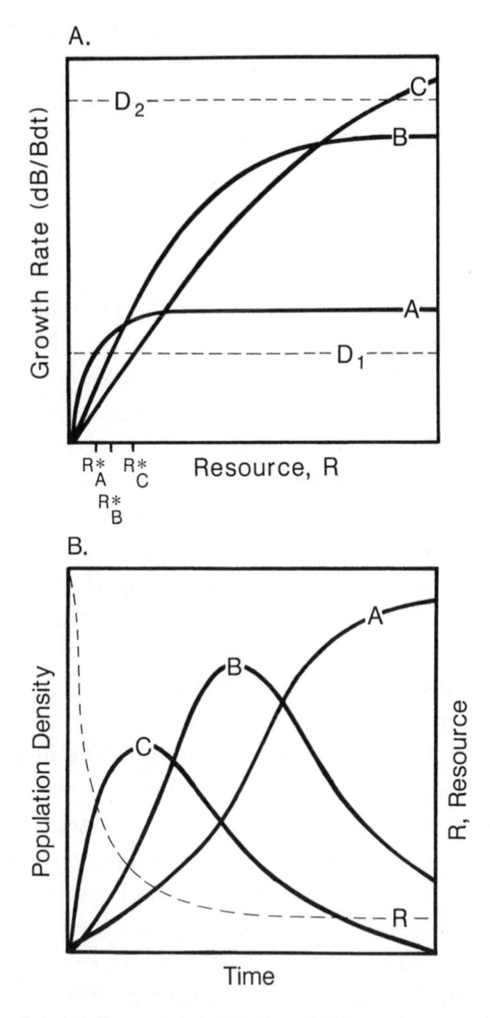

FIGURE 6.1. (A) Curves labeled *A*, *B*, and *C* show the dependence of growth rates (*dB*/*Bdt*, where *B* is plant biomass) for species *A*, *B*, and *C* on the availability of the resource, *R*. D_1 and D_2 are two different loss rates. The equilibrial resource requirements for the three species when growing in an environment with loss rate D_1 are shown by their R^*'s. (B) The qualitative dynamics of competition among these three species at loss rate D_1.

than species *A*. Thus, the dynamics of competition among these three species will show first a period of dominance by species *C*, then a period of dominance by species *B*, with species *A* eventually displacing species *B* and *C*. The amount of time required for species *A* to displace species *B* and *C* depends on initial densities, maximal growth rates, consumption rates of the limiting resource, the initial availability of the resource, and the dynamics of resource supply. For instance, imagine a situation in which all soil resources were available in excess and light was the only limiting resource. If species *C* were an annual herb, species *B* a perennial grass, and species *A* a tree, it could take many years for the much more slowly growing tree to create the closed canopy that would reduce the availability of light to the point at which the annual and herbaceous perennials were competitively excluded. Unfortunately, many of our observational and experimental studies occur over sufficiently short time spans that we may only see the periods of transient dominance by faster growing species, such as species *B* and *C*, and we may fail to see the eventual dominant.

The transient dominance by species *C* is caused by its maximal growth rate. If any environmental conditions or experimental manipulations caused there to be high resource availabilities, species *C* would be the initial dominant. Because it is consumption of limiting resources that causes resources to be held at low levels, manipulations, such as removals of one or more dominant species, that could cause resource levels to rise could lead to a period of transient dominance by species *C* or by other species with high maximal growth rates. Resource levels would be held at high levels by high loss rates, such as loss rate D_2, for which species C is the superior competitor. For disturbance rate D_2, species *C* would be the dominant without any period of transient dominance by an inferior competitor. Thus the dynamics of populations after an experimental manipulation or disturbance can include, but need not include, a period of transient dominance by species that are later displaced. Such transient dynamics mean that the "outcome" of such experiments is time-scale dependent. A well-replicated field experiment that ran for too short a time could yield a statistically highly significant "outcome" of competition that was more a measure of maximal growth rates than of competitive abilities for a limiting resource.

SIZE-STRUCTURED MODELS

In any competitive interaction that starts far from equilibrium, maximal growth rates are a major factor determining which species will become dominant initially. Maximal growth rates should increase with the proportion of photosynthate that is allocated to new photosynthetic machinery. For size-structured populations, such as terrestrial plants, the rate at which individuals increase in height might also be of critical importance when plants compete for light. During the initial stages of growth, those individuals that become taller faster will obtain a disproportionate share of the incident light, and thus suppress the growth of other individuals. If the plants that become taller, initially, are of shorter stature at maturity, this might cause a period of transient dominance during the initial phases of competition. Another factor that can influence the dynamics of plant competition is the dynamics of resources. Especially for size-structured populations, it is unlikely that soil nutrients and light would be reduced down to their equilibrium levels in a simple monotonic manner. This means that, dependent on starting conditions, resource levels could be driven along a complex trajectory that first favored one species then another, and so on, until resource levels approached an equilibrium, if that ever occurred. If resource levels never reached an equilibrium, one or more of the plant species could have sustained oscillations, limit cycles, or even chaotic dynamics that favored first one species and then another. I have seen all these dynamic patterns in simulations of resource competition in size-structured populations. In the following section I will present several of the types of dynamic responses shown by these models. These are illustrative of some of the possible dynamics of competition in size-structured plant populations, and are not meant to be a thorough treatment of the subject. I present them to illustrate that there need be no simple relation between the initial dynamics of plant competition and the ultimate pattern resulting from those competitive interactions. I also present them because they suggest an explanation for many of the short-term (1 to 10 years for old fields; 1 to 25 years for forests or light gaps) dynamic responses that comprise the bulk of published ecological observations. Clearly, as will be discussed in the following chapter, these models are just models of successional dynamics.

Multispecies Competition

Let us consider several cases in which 100 different species competed for a limiting soil resource and light, using the model ALLOCATE. All 100 species were identical in all ways except in their pattern of allocation to roots, leaves, and stems. The 100 species were chosen so as to have allocation patterns that spanned the full range of viable plants. For all cases, the loss rate was 0.2 wk^{-1}.

Let us first consider a case in which there was a nutrient-poor soil (TN = 100). Most of the 100 competing species never became dominant at any time. Indeed, 8 of the species accounted for 55% of the total plant mass at year 1 and for 92% by year 10. Averaged over the first 10 years, these 8 species accounted for 84% of the total plant mass. The dynamics of these 8 species, and that of the limiting soil nutrient and light penetration to the soil surface, are shown in Figure 6.2. The eventual competitive dominant, species *B*, had a long period of slowly increasing absolute abun-

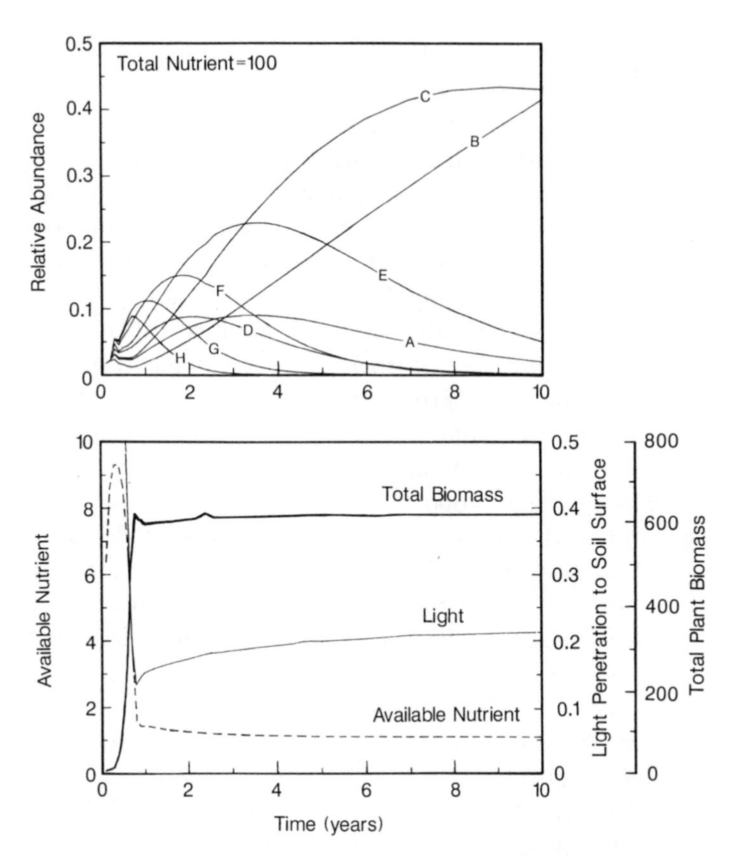

FIGURE 6.2. Population dynamics for a case in which 100 morphologically different, but otherwise identical, species compete on a nutrient-poor soil (TN = 100) at a loss rate of 0.2 wk^{-1}. Only 8 of the 100 species are shown here. These 8 species, labeled *A* through *H*, had allocation to leaf, stem and root, respectively, as follows. Species *A*: 0.55, 0.05, 0.40; species *B*: 0.50, 0.025, 0.475; species *C*: 0.55, 0.025, 0.425; species *D*: 0.60, 0.05, 0.35; species *E*: 0.60, 0.025, 0.375; species *F*: 0.65, 0.025, 0.325; species *G*: 0.70, 0.025, 0.275; species *H*: 0.75, 0.025, 0.225. These 8 species were the only species to attain dominance at any time during these simulations, which lasted for 40 years. In order to provide more detail of the initial dynamics, only the first 10 years are shown. By year 40, species *B* had displaced all other species.

dance as it monotonically approached its eventual equilibrium biomass. However, its relative abundance declined for several years before it began increasing. During the course of these competitive interactions, there were 5 other species that had a period of transient dominance followed by their competitive displacement. Thus, the dynamics of competition in this nutrient-poor habitat were fairly complex, and there was a considerable delay before the winning species achieved dominance. Most of the competitive displacement occurred during a period when total plant biomass (the sum of above and below ground biomass) was essentially constant. Light penetration to the soil surface initially was driven to about 0.13, and then increased slowly to about 0.22 as a greater proportion of plant biomass went belowground. As this occurred, available nutrient levels declined.

What morphological traits were favored at different times during these interactions? An easy way to visualize this is to show the allocation patterns of the species that achieved dominance at different times during the interaction (Fig. 6.3A). Of these 100 species, those that reached their peak abundance earliest, such as species H, were species with high allocation to leaves, and thus with large relative growth rates. These were subsequently replaced by species with progressively lower allocations to leaves but progressively higher allocation to roots (Fig. 6.3A). This process continued until species B achieved dominance. All eight of the species that ever attained dominance had low allocation to stems (indeed, the lowest used in these simulations, 2.5%). Thus, all of these species are basically rosette-like plants that increase in height very slowly. These simulations give results that are qualitatively similar to those of Figure 6.1. However, there is a major difference. There were many species with maximal growth rates greater than that of species H that never achieved dominance. These species allocated so little to roots that they were nutrient limited even initially, and did not achieve actual growth rates close to their maximal growth rates.

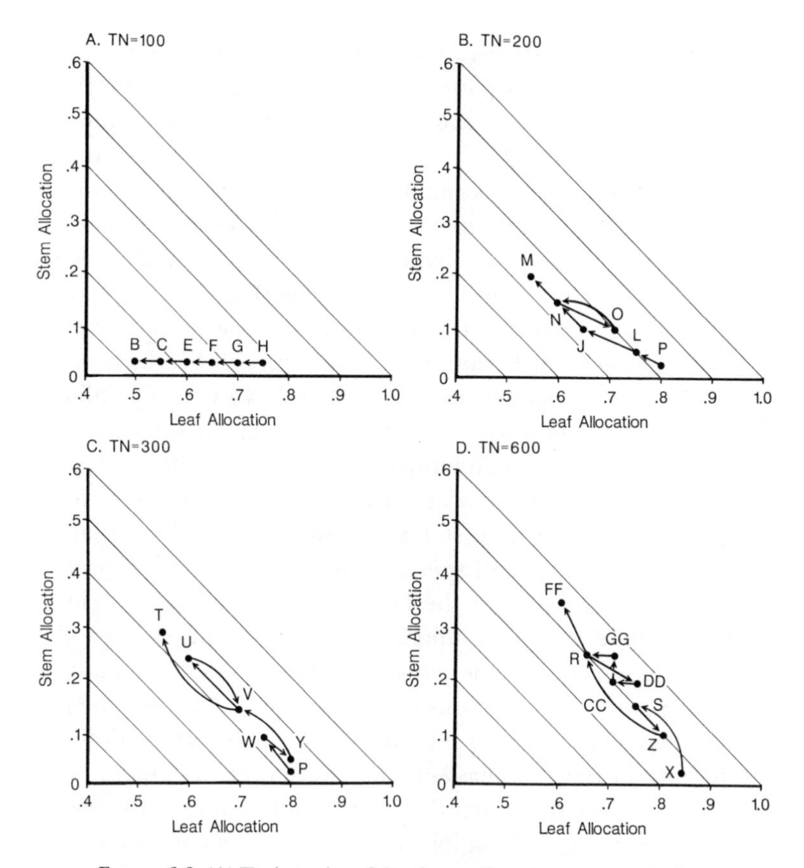

FIGURE 6.3. (A) Trajectories of dominant allocation patterns during the process of competitive displacement. Each dot shows the allocation pattern of the species dominant at a particular time. The letter associated with each dot is the name of that species. The arrows show the sequence of dominant allocation patterns during the processes of competitive displacement on this soil which had TN = 100. (B) Trajectories of dominant allocation patterns for TN = 200. (C) Trajectories for TN = 300. (D) Trajectories for TN = 600. Note that in some cases, the same species is dominant more than once. These trajectories correspond to the population dynamics illustrated in Figures 6.2, 6.4, 6.5, and 6.6. The same 100 species competed in all 4 cases. These 100 species were identical in every way except their allocation pattern. All results were based on numerical simulations using ALLOCATE. Note that axes do not range from 0 to 1.

Although a different series of species was dominant, a qualitatively similar pattern occurred when these same 100 species competed at this same loss rate, but on a more nutrient-rich soil (TN = 200; Fig. 6.4). The eventual competitive dominant, species *M*, increased slowly in absolute abundance through time, with its final absolute abundance having sustained oscillations. Its relative abundance declined the first year before increasing (Fig. 6.4A). Several periods of peak abundance of other species occurred before species *M* became dominant. Species *P*, which was the most abundant species at 0.5 year, has a very high allocation to leaves and a very low allocation to stems. The sequence of species attaining peak abundance after *P* represents species that allocate progressively less to leaves and progressively more to stems and roots (Fig. 6.3B). Again there is a sequence of species that attain a period of transient dominance, with the faster growing (leafier) species dominant earlier. The initial dominants on this richer soil were species with higher allocation to leaves than the initial dominants on poorer soils (compare Figs. 6.3A with 6.3B). Of the 100 species present, only those that had from 17.5% to 25% of their biomass in roots were ever dominant. Of these species, the initial dominants had high growth rates because of high allocation to leaves and low allocation to stems. The final dominant, species *M*, had a slower maximal growth rate because it allocated more to stem and less to leaves. Its greater allocation to stem allowed species *M* to more rapidly attain its maximal height and thus to gain a disproportionate share of the light compared to individuals of species that were leafier but allocated less to stem. Although its height at maturity was the same as that of all other species, at any given age individuals of species *M* were taller than individuals of the other dominant species. The transient dynamics observed in this case was thus a progressive replacement of morphologies from those that gained biomass fastest in high light habitats to those that grew

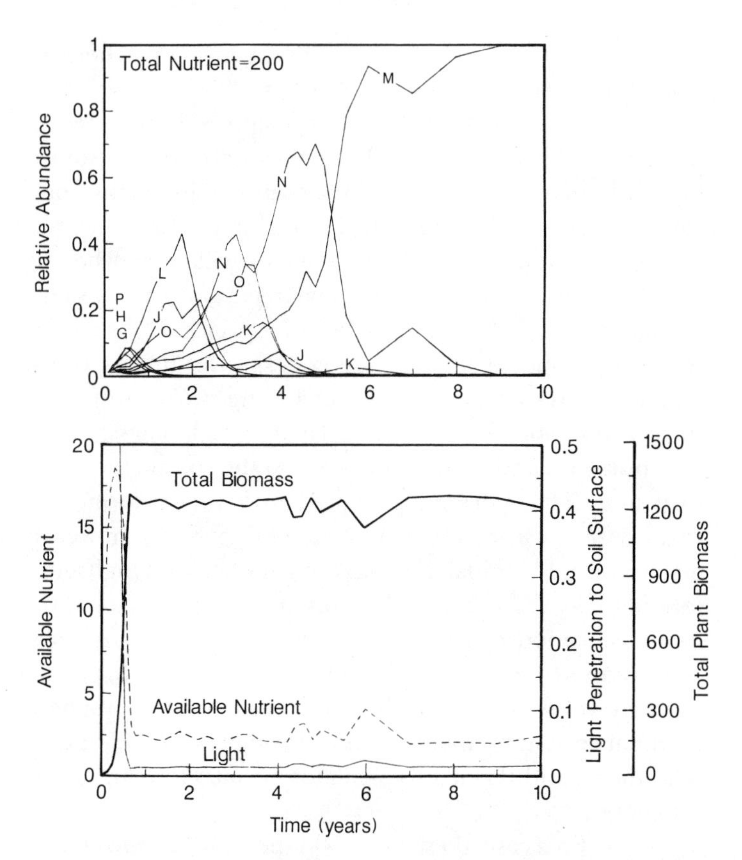

FIGURE 6.4. The dynamics of competitive displacement for the same 100 species used for Figure 6.2, except that here the soil nutrient supply rates are higher (TN = 200). Only the 10 species that ever attained dominance are shown. These 10 species, labeled *G* through *P*, had allocation to leaf, stem, and root, respectively, as follows. Species *G*: 0.70, 0.025, 0.275; species *H*: 0.75, 0.025, 0.225; species *I*: 0.60, 0.20, 0.20; species *J*: 0.65, 0.10, 0.25; species *K*: 0.65, 0.15, 0.20; species *L*: 0.75, 0.05, 0.20; species *M*: 0.55, 0.20, 0.25; species *N*: 0.60, 0.15, 0.25; species *O*: 0.70, 0.10, 0.20; species *P*: 0.80, 0.025, 0.175.

fastest in low light habitats because of their greater height, given the constraint of a soil with TN = 200.

When these same 100 species competed in a more nutrient-rich habitat (loss = 0.2; TN = 300), both the absolute and relative abundance of the eventual dominant, species *T*, decreased for four years before increasing (Fig. 6.5). Its absolute abundance had sustained oscillations when it was dominant. There were six major periods of transient dominance by other species before species *T* displaced them. Most species had oscillatory responses during competition, with species *U* and *V* each having two or three distinct peaks about 2 years apart. Total plant biomass, available nutrient, and light penetration to the soil surface also oscillated. The morphologies of the dominant species at various times during these interactions (Fig. 6.3C) showed a general trend of initial dominance by species with high allocation to leaves followed by dominance by species with lower allocation to leaves but higher allocation to stems. Allocation to roots was fairly constant during this progression. However, as already discussed, some species attained dominance at more than one time during the period of displacement, and this led to oscillatory trends in plant morphologies.

For the final case, these same 100 species were allowed to compete in a very nutrient-rich habitat (TN = 600). Both the absolute and relative abundance of the eventual dominant, species *FF*, decreased for the first 2 years and then it started a slow, unsteady increase toward total dominance (Fig. 6.6). The first species to attain dominance, species *X*, had high allocation to leaves (85%) and low allocation to stem (Fig. 6.6 and Fig. 6.3D). The series of species that had periods of transient dominance following it had increasingly great allocation to stem and increasingly lower allocation to leaves (Fig. 6.3D). They also had lower allocations to roots than the initial dominants. All species tended to oscillate, and one species, *R*, was the most abundant species at

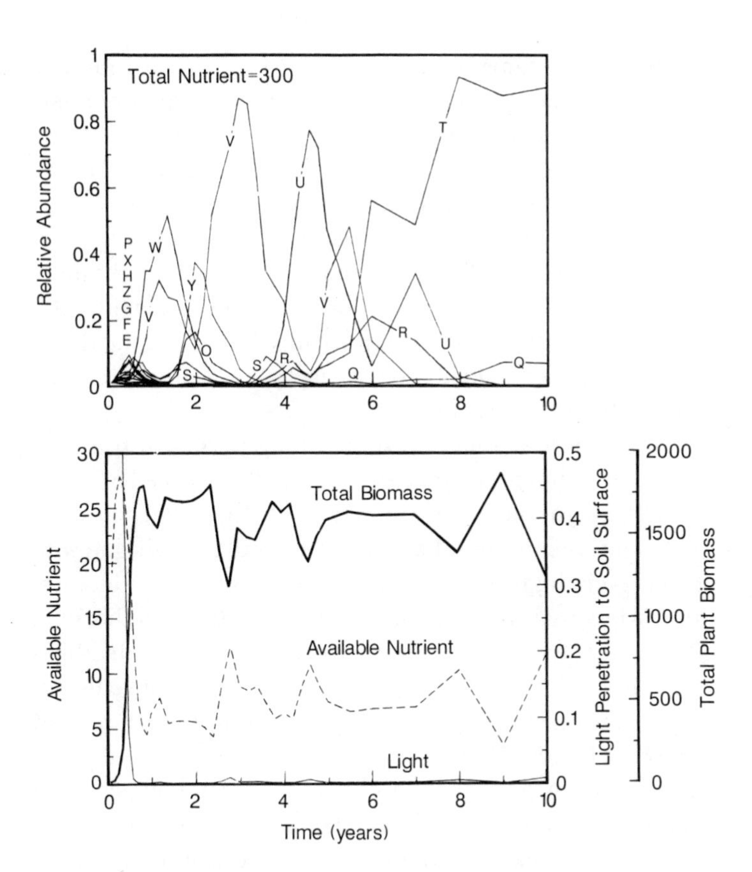

FIGURE 6.5. The dynamics of competitive displacement for the 100 species used in Figures 6.2 and 6.4, except that here the soil nutrient supply rate is higher (TN = 300). Only the 16 species that ever attained dominance are shown. The ten species labeled *Q* through Z had allocation to leaf, stem, and root, respectively, as follows. Species *Q*: 0.60, 0.30, 0.10; species *R*: 0.65, 0.25, 0.10; species *S*: 0.75, 0.15, 0.10; species *T*: 0.55, 0.30, 0.15; species *U*: 0.60, 0.25, 0.15; species *V*: 0.70, 0.15, 0.15; species W: 0.75, 0.10, 0.15; species *X*: 0.85, 0.025, 0.125; species *Y*: 0.80, 0.05, 0.15; species *Z*: 0.80, 0.10, 0.10. Allocations for the other six species are given in the preceding legends.

two different times. These oscillations are clearly indicated by the oscillatory nature of the trajectory of the dominant morphologies (Fig. 6.3D). Total biomass, nutrient availability, and light penetration oscillated throughout the simulations. As for the previous two cases, the eventual dominants displaced the other species because their greater allocation to stem allowed them to increase in height more rapidly and thus outcompete other species for light. This trait was favored even though it caused the eventual dominant to have a maximal growth rate that was much less than that of the initial dominant.

All of these simulations show a long period of transient dominance by species that are eventually displaced by a species that is the superior competitor for the particular environmental conditions. Each of these simulations began with a relatively high seed density, but with total plant biomass much less than the eventual level attained at equilibrium. Thus, at the beginning of each simulation the individuals were living in a relatively high light and high nutrient situation. This favored species that allocated little to stems and much to leaves. As increased plant biomass reduced available nutrient and light levels, other morphologies were favored. However, the identities of the species that had periods of transient dominance were different for each habitat (Fig. 6.3). At equilibrium, the eventual dominants for a habitat with a loss rate of $D = 0.2\ \text{wk}^{-1}$ are species with allocation patterns that fall along the curve of Figure 4.5. Figure 6.3 provides an important contrast to Figure 4.5: of all the possible allocation patterns, the only species ever to attain a period of transient dominance were those that allocated more to leaves than the eventual equilibrial dominants. There are many viable species that have lower allocation to leaves than the eventual competitive dominants (see Fig. 4.4), but these species never attained a period of transient dominance in any of the simulations. The trajectories of Figure 6.3 suggest that almost the full range of

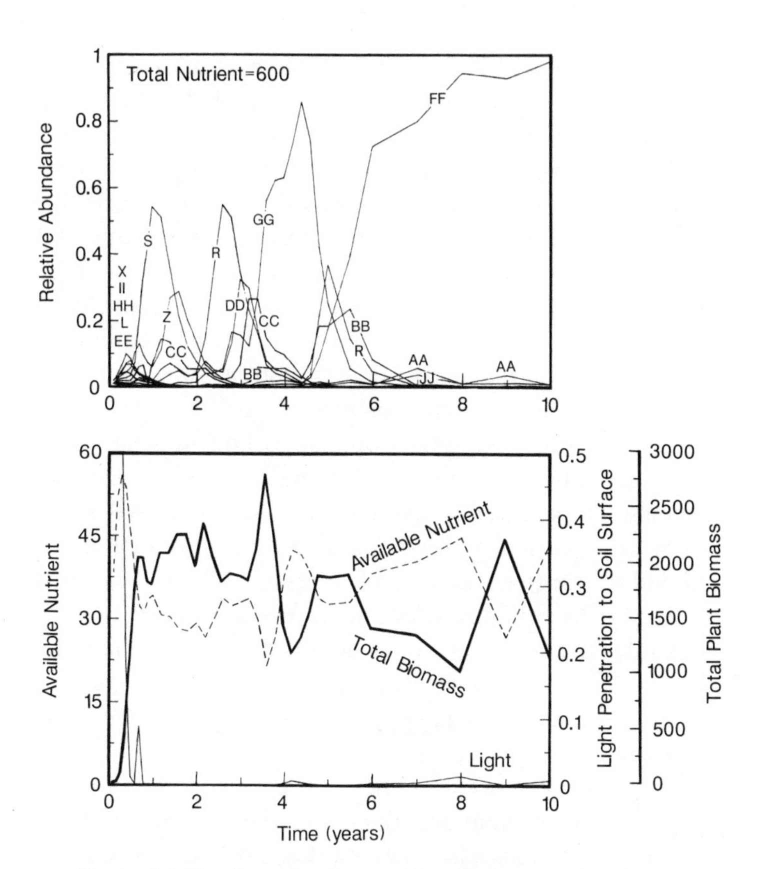

FIGURE 6.6. The dynamics of competitive displacement for the same 100 species used for Figures 6.2, 6.4, and 6.5, except that here the soil nutrient supply rate is higher (TN = 600). Only the 15 species that ever attained dominance are shown. The species labeled *AA–JJ* have allocations to leaves, stems, and roots, respectively, as follows. Species *AA*: 0.55, 0.40, 0.05; species *BB*: 0.65, 0.30, 0.05; species *CC*: 0.70, 0.20, 0.10; species *DD*: 0.75, 0.20, 0.05; species *EE*: 0.50, 0.45, 0.05; species *FF*: 0.60, 0.35, 0.05; species *GG*: 0.70, 0.25, 0.05; species *HH*: 0.90, 0.025, 0.075; species *II*: 0.85, 0.05, 0.10; species *JJ*: 0.65, 0.325, 0.025. Allocation patterns for the other species are given in the legends to the preceding figures.

morphologies with greater leaf allocation than the eventual competitive dominants could attain a period of transient dominance at some time on the appropriate soil.

There is an interesting comparison between this pattern and the results shown in Figures 4.8 and 4.9. Figures 4.8 and 4.9 show that, at equilibrium, there are environmental conditions that favor the full range of viable morphology patterns. Figure 6.3 shows that a wide range of morphologies are also favored by non-equilibrium conditions. Indeed, the species that attain a period of transient dominance in recently disturbed habitats are also the species that are the eventual equilibrial dominants of habitats with high loss rates. This pattern goes even one step further. For any given soil type (i.e., any given TN value in the model ALLOCATE), the sequence of species (morphologies) attaining a period of transient dominance tends to go from those that would be the equilibrial dominants on that soil type at high loss rates to those that would be the equilibrial dominants on that soil at progressively lower loss rates. This sequence continues until the actual equilibrial loss rate of the habitat is reached.

This similarity between the transient dynamics of competitive displacement and the equilibrial outcome of competition in habitats with different loss rates occurs because of the patterns of resource availability. For any given soil type, habitats with high loss rates have higher levels of the limiting soil nutrient, higher light intensity at the soil surface, a lower canopy and thus a less steep vertical light gradient. Similar conditions exist early during the process of competitive displacement. The low plant biomass and shortness of the plants early during population growth means that there are high levels of limiting soil nutrients, high light intensity at the soil surface, and a low canopy. Thus, the resource levels that exist during the early phase of growth tend to favor the same traits that allow species to be dominant in habitats with high loss rates. As populations grow, the equilibrial level of the limiting soil resource and the equilibrial vertical light gradient are approached, which favors species with morphologies different from those of the initial dominants.

The transient dynamics seen within a habitat are much like the dynamics of succession (Chapter 7). Such transient dynamics are strongly influenced by maximal growth rates and resource availabilities. Because maximal growth rates are determined by allocation patterns, and because different allocation patterns are favored in different habitats, transient dynamics should be a ubiquitous feature of terrestrial vegetation. Any disturbance or experimental manipulation that reduces plant densities or otherwise increases resource availabilities above equilibrium levels should lead to a successional sequence—i.e., to a period of transient dominance. Just such transient dynamics were observed in the Park Grass fertilization experiments, performed at Rothamsted, England, over the past 130 years. In two cases in which all nutrients were provided, with nitrogen added as nitrate, the eventual dominants were *Alopecurus pratensis* and *Arrhenatherum avenaceum* (Fig. 6.7 B and C). *Alopecurus* increased in absolute abundance throughout the experiment, but the absolute abundance of *Arrhenatherum* declined by 7× and 100× for the first 14 years in both plots receiving nitrate. It was only after this decline that it began a slow increase in abundance. It did not have comparable declines or increases in abundance in the unfertilized control plots (Fig. 6.7A) or in the plots receiving nitrogen as ammonium (Fig. 6.7 D-F). *Agrostis* and *Festuca* were stable or increased in abundance for many years before declining (Fig. 6.7 B-C). The plots receiving complete mineral fertilizer with nitrogen as ammonium also had periods of transient dominance. For instance, one plot was dominated first by *Dactylis* and then by *Agrostis* before the eventual dominant, *Holcus lanatus*, displaced them (Fig. 6.7E). There was a similar succession of dominant species in a plot that

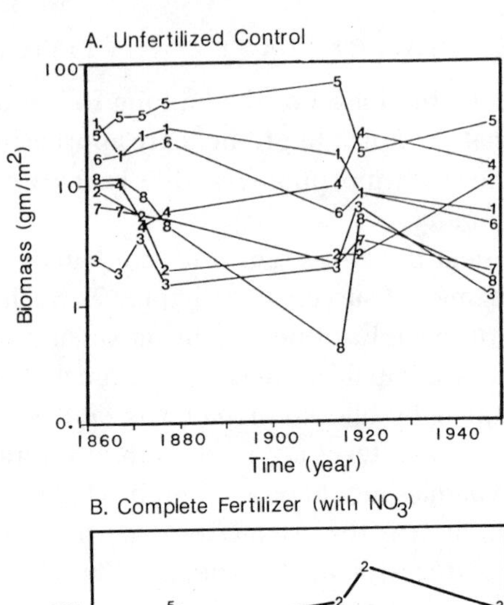
A. Unfertilized Control
Biomass (gm/m2)
100
10
1
0.1
1860
1880
1900
1920
1940
Time (year)

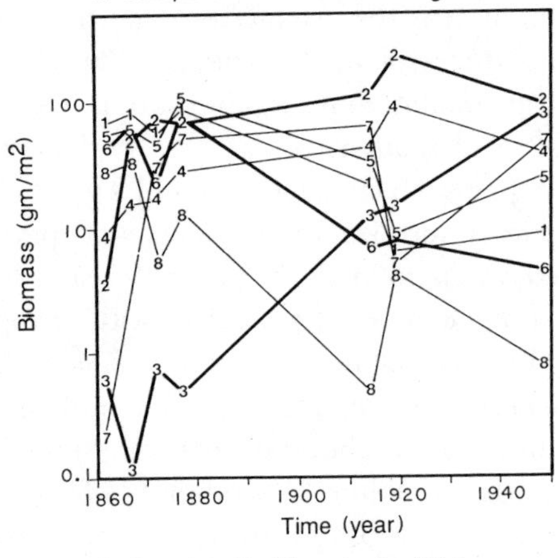
B. Complete Fertilizer (with NO3)
Biomass (gm/m2)
100
10
1
0.1
1860
1880
1900
1920
1940
Time (year)

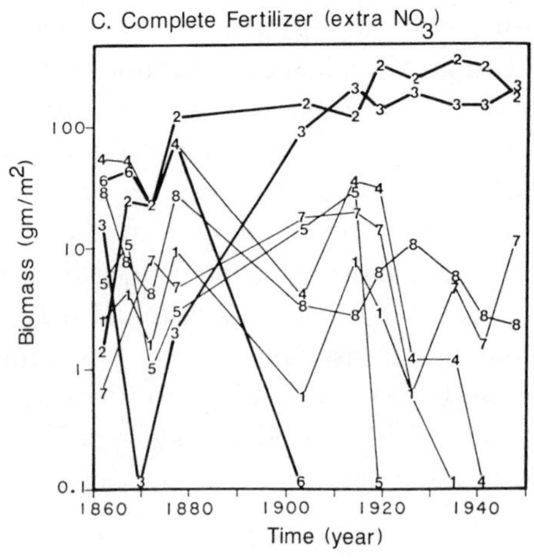
C. Complete Fertilizer (extra NO3)
Biomass (gm/m2)
100
10
1
0.1
1860
1880
1900
1920
1940
Time (year)

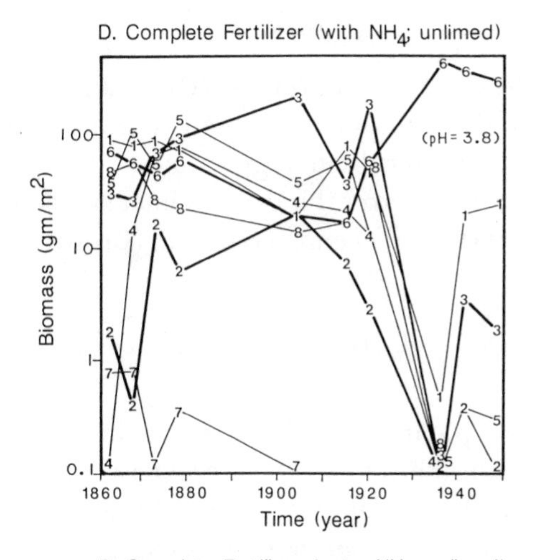

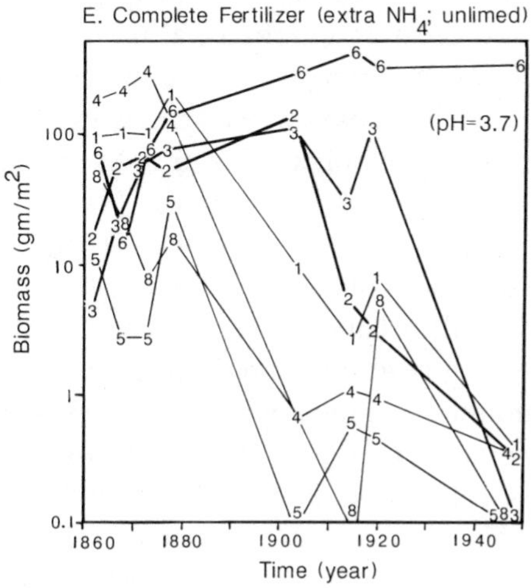

1 = *Agrostis*	5 = *Festuca*
2 = *Alopecurus*	6 = *Holcus*
3 = *Arrhenatherum*	7 = *Lathyrus*
4 = *Dactylis*	8 = *Rumex*

FIGURE 6.7. Dynamics of major plant species during the first 100 years of the Park Grass Experiments at Rothamsted Experimental Station, England. Note the periods of transient dominance by species that are later displaced. Some of the transient dynamics may be caused by slow changes in soil pH associated with particular fertilizers. Biomass is shown on a log scale. Figure modified from Tilman (1982).

received less ammonium (Fig. 6.7D). Transient dynamics have been observed in many other experiments, including the Cedar Creek fertilization experiments (Chapter 8, Figs. 8.23–8.26), forest trenching experiments (Toumey and Kienholz 1931), and mammal removals (Brown et al. 1986).

TRANSIENT DYNAMICS AND ECOLOGICAL UNDERSTANDING

Transient population dynamics may be a major factor slowing down the rate at which ecologists gain an understanding of the workings of nature. Because of the tendency to extrapolate from short-term experiments, our field may place too great an emphasis, at the present time, on traits of species that may be of little direct importance in determining the habitats in which species are dominant and which may have been only weakly or indirectly selected for in the evolutionary past of a species. There is still much research being performed in ecology that looks at the immediate responses of organisms to experimental manipulation (i.e., a disturbance) and assumes that these immediate responses give insight into the forces structuring equilibrial communities. This approach, however, is likely to give very inappropriate "answers." Consider someone studying the role of a limiting soil nutrient in structuring vegetation. The obvious way to determine the effect of the limiting soil nutrient would be to establish a well-replicated field experiment in which the nutrient was added (and subtracted, if possible) at different rates in different treatments. After a few years, the experiment might yield highly statistically significant changes in the species composition of the treatments. To determine if the limiting soil nutrient is the cause of the natural vegetation patterns in this habitat, the usual approach would be to determine if the species abundances along the experimental gradient were similar to those on a natural, "equilibrial" gradient in the field. Is this, though, a good test of the role of the nutrient in structuring this community? Consider the case illustrated in Figure 6.8. Figure 6.8A shows which morphologies are predicted to be competitively superior at equilibrium in a habitat with a particular loss rate (based on the model ALLOCATE). The logic presented above would suggest that enriching a habitat with a TN = 200 soil so as to have it become a TN = 500 soil should lead to the eventual dominance by a species with the morphology at that point on the curve of Figure 6.8A. However, the trajectory that this community takes after nutrient enrichment is not simply one of moving from the initial dominant along the curve of Figure 6.8A to the eventual dominant. Rather, nutrient addition leads to a period of transient dominance by species that do not fall on this curve (Fig. 6.8B). In this simulation, it took about 15 years for the equilibrial dominant species to displace the various transiently dominant species. Moreover, the model predicts that the sequence of transient dominants is a regular, repeatable feature of the dynamics of competition. As such, it is quite likely that the well-replicated experiment would yield highly statistically significant results after one year showing that nutrient addition favored one species. There could be equally significant results after two years showing that a different species was favored, and so on until the equilibrial dominant had displaced the other species. There would be many years during which the results could be easily misinterpreted by someone who did not realize that 15 or more years were required to determine the long-term effect of the manipulation. An even longer period would be required for experiments performed in forested vegetation.

It is for this reason that I believe that much of the work on light gaps in forests, on small-scale disturbances in grasslands, and on regeneration strategies in general may contain an element of myopia. Many such studies have focused on the short-term transient dynamics of such patches, and tried to extrapolate this information to explain the structure of the entire community. However, such short-term studies give too great an importance to differences in maximal growth rates and tend to obscure the underlying differences in competitive abilities for limiting soil resources and light that may be the important forces structuring the communities. The differences in maximal growth rates that seem so important in short-term studies may only be an evolutionarily indirect effect of selection for different morphologies, and may have little direct effect on the equilibrial structure of the community. The responses of seedlings and saplings to the average conditions imposed by the canopy trees may be as important as their responses to disturbances. Such disturbances are surely an important factor maintaining diversity in a forest, and forests with different long-term average disturbance rates should have different compositions. However, these compositional shifts should be explained by the underlying mechanisms. High rates of disturbance favor species with different physiologies and morphologies than do low rates of disturbance, because high rates of disturbance change average resource availabilities.

One of the more frustrating aspects of doing ecological research is the patience that is required. Patience is required because human lives, careers, ambitions, and curiosity are often shorter than the time required for a community to respond fully to experimental manipulations or natural disturbances. Thus, it is just such periods of transient dominance that are studied in most ecological experiments on terrestrial plant communities. Unfortunately, the traits of the initial dominants often tell us little if anything about the factors that structure the natural community. Nor is it a trivial matter to use the initial population dynamics of the species to determine which will be the eventual winner. Figures 6.5 and 6.6 showed that, on a rich soil, the eventual competitive dominants may decrease in their relative abun-

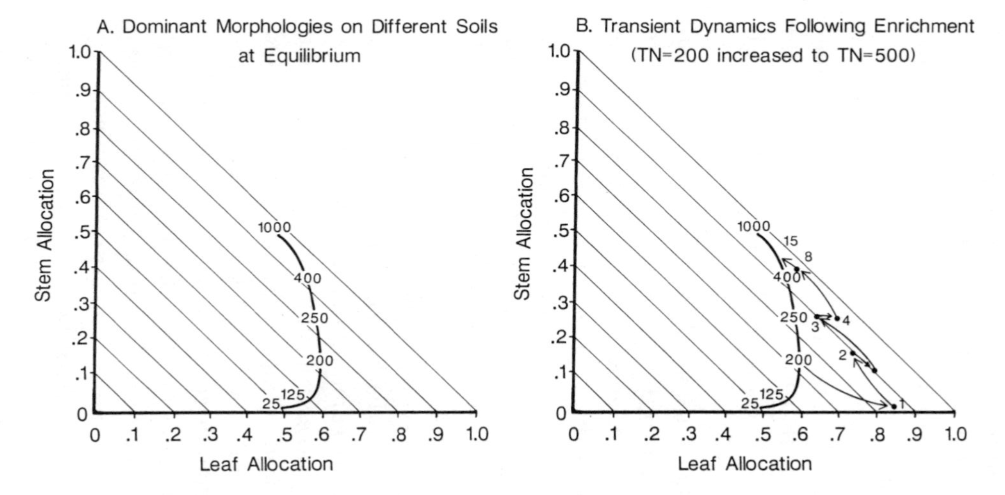

FIGURE 6.8. (A) At equilibrium, in habitats that have identical, and constant, loss rates, the nutrient supply rate, TN, determines which morphologies will be competitively superior. This shows the competitively superior morphologies for habitats ranging from those with TN = 25 to those with with TN = 1000. (B) If a habitat with a TN = 200 soil were experimentally manipulated to make the soil be a TN = 500 soil, there would be a sequence of species that attained transient dominance. These species would not fall along the equilibrial curve of part A of this figure, but would be biased toward species with higher leaf allocation. The smaller numbers (1, 2, 3, 4, 8, and 15) are the number of years the model ALLOCATE predicted would be required for each of these different morphologies to attain its period of transient dominance. Note that this case started with high biomass of the species that is the equilibrial dominant of TN = 200 habitats, and low biomass of all other viable allocation patterns.

dance (and even absolute abundance) for a long time before increasing. For instance, species *FF* of Figure 6.6 initially decreased in abundance and did not start to increase until the other species had formed a dense tall stand in which it had a competitive advantage because of its greater allocation to stem. I have not found any way to use the early (first 2 or 3 years) dynamics of competition (as in Figures 6.2, 6.4, 6.5, and 6.6) to predict which species would be the eventual dominants for a given soil type or loss rate. Most field competition experiments that have been performed, though, have lasted just one or two years (Schoener 1983; Connell 1983).

If these simulations are at all indicative of the mechanisms of competition in terrestrial plant communities, these results might be taken to suggest that inferences about the dependence of competitive abilities on habitat characteristics could best be made either from direct observation in mature, equilibrium communities or from long-term experiments. Either of these approaches might be preferable to the short-term experiments that have become so fashionable, of late, in ecology. Both of these approaches, though, have major drawbacks. Correlational studies based on patterns in mature communities are just that—correlations. They are not necessarily indicative of causation, and thus could easily lead to misinterpretation of cause and effect relations. This is especially true in ecology because of the many direct and indirect pathways whereby one species may affect another (Levine 1976; Lawlor 1979; Vandermeer 1980). Indirect effects also make it difficult to use a single long-term experiment to interpret unambiguously the mechanisms of interspecific interactions in multi-trophic-level systems. A nutrient addition experiment, for instance, could lead to the eventual dominance of a particular species not because of that species' competitive abilities but because that species was most resistant to an herbivore that increased in abundance as productivity increased following nutrient addition. Because of such difficulties, it is unlikely that ecologists will ever be able to devise a single, critical experiment that, when performed in natural communities, could unambiguously distinguish between various competing hypotheses. However, correlational studies in natural communities, long and short-term experimental manipulations of natural communities, and studies of the mechanisms of interspecific interaction under controlled experimental conditions can all provide critical pieces of information in helping us understand how nature functions. It will take a multifaceted approach—one that synthesizes a wide variety of experimental, observational, and theoretical approaches—to develop a predictive understanding of the natural world.

I see little hope of using purely phenomenological approaches as a shortcut. The two major approaches to plant competition, in recent years, have both have been phenomenological: (1) field density manipulation experiments, such as removal experiments or transplant experiments (see reviews in Schoener 1983 and Connell 1983) and (2) the de Wit replacement series. There are major conceptual problems with both approaches. The density manipulation approach is based on the Lotka-Volterra-derived definition of pairwise competition—that two species compete when an increase in the density of one leads to a decrease in the density (or growth rate) of the other, and vice versa. This definition, however, applies only to an interaction between two species in the absence of any other species. As soon as there are more than two species, indirect effects mediated through other species can greatly modify the total effect of one species on another. As Lawlor (1979) demonstrated, two species whose only mechanism of direct interaction is competition, can function, in a multi-species community, as mutualists. Density manipulation experiments, as they have been performed and interpreted to date, would detect this apparent mutualism, not the direct competition (Tilman 1987b). Fowler (1981) found just such apparent mutualism in a series of removal experiments performed on a North Carolina grassland, and explained the apparent mutualism as being an indirect effect of multi-species competition. Bender, Case, and Gilpin (1984) showed that *every* species has to be removed in separate, replicated experiments to determine unambiguously whether or not a given pair of species are direct competitors. Alternatively, Bender, Case, and Gilpin (1984) suggest that experiments be designed to determine the immediate transient effect of a change in the density of one species on the densities and growth rates of all species. With this approach, only two replicated experiments are required to determine how a pair of species interact, assuming that all interactions are actually determined by direct density effects, with no time lags, as assumed in the Lotka-Volterra model. Such experiments, though, would fail to show competition between two species that were competing for the same resource if resource availabilities changed slowly in response to a density manipulation, if there were a time lag in the response of growth and death to resource levels, or if resource dynamics were complex, as in Figures 6.5 and 6.6. Thus, density manipulation experiments, especially short-term density manipulation experiments, are of limited utility for describing or explaining competition in nature. However, the approach of Bender et al. (1984) is important because it illustrates that short-term experiments can be useful if interpreted within the context of the appropriate dynamic model.

The replacement series approach of de Wit (1960) is logically flawed, and thus incapable of predicting the long-term outcome of interspecific competition. Inouye and Schaffer (1981) demonstrated that the predictions of the de Wit approach depended on the total population density used in an experiment. Spitters (1983) and Connolly (1986) have also demonstrated that a replacement series approach is incapable, in theory, of predicting the outcome of interspecific competition because it ignores the dependence of competitive effects on total density. Furthermore, the de Wit approach attempts to use initial dynamics to predict the long-term outcome of competition. However, as already discussed, the initial dynamics of competition need have little, if any, relation to the long-term outcome of competition.

Thus, the two major approaches that have been used to describe the phenomenon of interspecific competition among plants must be viewed with extreme caution. Our best hope of understanding nature, I believe, is to study the mechanisms of interspecific interactions, the evolutionary basis for the selective forces that molded these interactions, and the relationships between these and the major physical and biotic constraints of the habitat.

In doing such work it should be remembered that differences in maximal growth rates, the presence of indirect effects, the existence of age structure, and a host of other factors could all cause natural communities to have a period of transient dynamics following a manipulation. Unless an experiment is performed for a sufficiently long period that the equilibrium outcome is observed, it would be quite possible to obtain highly statistically significant experimental results that bore little relationship to the ideas the experiments were testing. Short-term experiments are useful, but only as tests of the explicit, short-term predictions of dynamic models.

SUMMARY

The dynamics of competition can be complex, with many species attaining periods of transient dominance before being displaced (Figs. 6.2–6.6). There is, though, a definite pattern to such transient dynamics. The initial dominants are shorter species with higher maximal growth rates, i.e., higher allocation to leaves and lower allocation to stems. These are replaced by a series of species that have progressively lower maximal growth rates because of increasingly higher allocations to roots and/or stems. On nutrient-poor soils, the initial dominants are species that are the equilibrial competitive winners at high loss rates on poor soils (compare Fig. 6.3 with Fig. 4.18). These are replaced by species that are also competitive dominants, at equilibrium, on similarly nutrient-poor soils, but at progressively lower loss rates. Thus, the progression of transient dominants on a nutrient-poor soil represents a series of species that tend to be the superior competitors on such soils, at equilibrium, but at progressively lower loss rates. A similar pattern of progressive replacements occurs on rich soils. The sequence of species attaining a period of transient dominance on any given soil is similar to the sequence of species that would be seen, at equilibrium, along a gradient from high to low loss rates on this same soil. Although there is not a perfect mapping of the equilibria associated with different loss rates into the transient dynamics of competition, the similarities suggest that the same process that leads to the eventual dominance of particular species at different rates of nutrient supply and different loss rates also influences the short-term dynamics of competition. This process is resource competition.

The initial dynamics seen during competition may often be transient dynamics that are not directly or simply related to the eventual outcome of competition. Because these dynamics often occur on the time scale of grants and within the attention span of many ecologists, much of the "common wisdom" of ecology may be based on transient dynamics that are not directly related to the underlying selective pressures that have molded species' life histories and the major constraints that have shaped natural communities.

CHAPTER SEVEN

Succession

Terrestrial plant succession, generally defined as the dynamics of plant populations on an initially bare substrate, has been a subject of ecological study and debate since the turn of the century (Cowles 1899; Cooper 1913; Clements 1916; Gleason 1917, 1927). Despite the large number of papers and books that have been written on the subject, there is still no consensus among ecologists as to the causes of successional patterns. Clearly, this is partly because succession is such a broad subject that it is extremely unlikely that any single factor or process will ever explain all successions. Each species has physiological, morphological, and life history characteristics that are unique to it. Each habitat has a unique substrate, geomorphology, climate, and past history. The initial densities of colonists and the probabilities of colonization by various species are unique to each successional event. Further, because colonization is necessarily a probabilistic event and because successful colonists can have great potential for rapid growth in a previously vacant habitat, succession is unavoidably stochastic. Nevertheless, succession is an often repeatable process locally that shares many features from habitat to habitat worldwide (see papers in West, Shugart, and Botkin 1981). It is these similarities that have kindled an interest in succession. It is the differences among successions, however, that may provide the best clues to its causes.

PRIMARY SUCCESSION

By definition, successions start with a bare substrate. For primary successions, such as those occurring after glacial recession, sand dune formation, landslides, volcanic eruption, or major erosion, the bare mineral substrate is initially devoid of organic humus and plant propagules. For secondary successions, such as those occurring after clearcutting or farming, an organic soil and propagules of various plant species are present at the start. Most successions fall somewhere between these two idealized extremes. What forces cause the succession of species seen on these substrates?

There are many possibilities, none of which are mutually exclusive. I will start by considering the possible role of limiting resources in determining the pattern of primary succession. Later I will consider the role of resources in secondary succession and then I will discuss the role of other processes in both primary and secondary succession. Except for nitrogen, all of the mineral elements required by plants occur in the parent material in which most soils form (Jenny 1980). Thus, after glacial recession or sand dune formation in mesic habitats, the major soil factor limiting plant growth is nitrogen. In Glacier Bay, Alaska, for instance, there has been a rapid recession of glaciers during the 230 years since the end of the Little Ice Age. This has provided a dramatic chronosequence that documents the pattern of primary succession (Cooper 1923, 1939; Crocker and Major 1955; Lawrence 1958). Many of the later dominants of succession at Glacier Bay, such as cottonwoods and spruce, were among the initial colonists of the newly deposited glacial till, but these often did not survive, or they grew in a chlorotic, stunted form indicative of extreme limitation by nitrogen (Lawrence 1979). Most of the early dominants were capable of nitrogen fixation (Lawrence 1979). For instance, some areas were initially dominated by a "black crust" of nitrogen-fixing cyanobacteria (Worley 1973). Other initial dominants capable of nitrogen fixation included lichens such as *Stereocaulon* and *Lempholemma*, a rose (*Dryas drummondii*), the alder *Alnus crispa*, various lupines, and other legumes (Lawrence et al. 1967; Reiners et al. 1971; Lawrence 1979). As these early dominants lost leaves or roots, and as they died, the nitrogen-containing organic matter they produced became available to various decomposer species. These decomposers retained much of this nitrogen as well as nitrogen provided from atmospheric sources, causing total soil nitrogen levels to increase. The decomposer species also excreted nitrogenous wastes, some of which were then available to vascular plants. Thus, the early dominance by nitrogen-fixing organisms, as well as the accumulation of nitrogen provided from various atmospheric sources, led to a gradual increase in total soil nitrogen levels at Glacier Bay. Crocker and Major (1955) observed that total soil nitrogen levels increased from 5 to 10-fold within the first 100 years of succession at Glacier Bay, based on a chronosequence of successional sites on matched parent material. Crocker and Dickson (1957) observed similar increases in total soil nitrogen in several other chronosequences on Alaskan glaciers. Olson (1958) observed a similar pattern of increase in total soil nitrogen in the sand dunes of southern Lake Michigan. Such increases in total soil nitrogen, at least over this great a range, should correspond with increased rates of nitrogen supply. The rate at which the mineral forms of nitrogen (NH_4 and NO_3) are made available in a soil is called the nitrogen mineralization rate of the soil. Nitrogen mineralization rates are known to depend on many factors, such as soil pH, soil oxygenation, moisture, temperature, organic matter quality, carbon to nitrogen ratios, and total soil nitrogen (Chichester et al. 1975; Melillo et al. 1982; Vitousek et al. 1982; Pastor et al. 1982). Robertson and Vitousek (1981) and Robertson (1982) found that the mineralization potential of the sand dune soils of southern Lake Michigan increased with their total nitrogen content, and thus with their age. This means that during the first 100 or 200 years of primary succession, a bare mineral substrate that is extremely nitrogen deficient becomes an increasingly nitrogen-rich organic soil with a higher rate of nitrogen supply.

As these soils became more nitrogen rich, total plant biomass increased. For instance, at Muir Inlet of Glacier Bay, the initial dominants were soil algae and lichens, followed by the nitrogen-fixing shrub *Dryas drummondii*. These were then displaced by alders, which grew to heights of 3–10 m, and which were displaced by cottonwoods that grew up in their shade and then overtopped them. Lawrence (1979, p. 217) stated that "By the time 50 years have passed since the melting of the ice, a dense alder thicket up to 10 m (33 ft) tall had developed at Muir Point from which have emerged the same cottonwood trees that became established soon after ice recession, but which could not grow erect, at some places for over 40 years, until the alder provided them with adequate amounts of nitrogen for rapid erect growth." These cottonwoods were then replaced by even taller spruce and hemlock, which were able to grow up through the shade cast by the cottonwoods. A qualitatively similar pattern occurred in the sand dunes of Lake Michigan (Cowles 1899; Olson 1958), except that few of the initial dominants were nitrogen fixers (Olson 1958). Olson suggests that the accumulation of organic nitrogen in the sand dune soils could be explained by normal atmospheric nitrogen inputs coupled with low loss rates of nitrogen. The dunes started as a low nitrogen but high light habitat. As soil nitrogen increased, total plant biomass increased, and plants of taller stature were favored. In some areas, marram grass and other "dune stabilizing grasses" were the initial dominants, followed by a variety of forbs, then shrubs, and finally jack pine, white pine, and black oak. In areas with a different aspect and slope, and perhaps with higher clay content (Olson 1958), the early successional grasses were followed by forbs, shrubs, basswood, red oak and then sugar maple. Olson stressed that there was no single endpoint to this succession, but that the species composition eventually attained on a sand dune depended on its slope, aspect, and original parent material.

These studies suggest certain features that should apply to a wide variety of primary successions. At the start of primary succession, the habitat experienced by the seedlings of new colonists has low nutrient but high light availability. As various soil-forming processes, mediated by numerous decomposer species, increase the availability of the limiting soil resource or resources, total plant biomass increases and light penetration to the soil surface decreases. Thus, primary successions, at least during their first few hundred years, can be thought of as a gradient through time in the relative availability of one or more limiting soil resources and light at the soil surface. If the rate of accumulation of the limiting soil resource is slow relative to the rate of competitive displacement, many of the features of primary succession might be explained as a slowly shifting trajectory of equilibrial plant communities, with the composition at any point mainly determined by the relative availability of the limiting soil resource and light. I have called this the resource ratio hypothesis of succession (Tilman 1982, 1985). However, as will be discussed later in this chapter, transient dynamics could also influence primary successions.

The resource ratio hypothesis may be easily illustrated using resource-dependent growth isoclines (Fig. 7.1). The five species labeled *A* through *E* are differentiated in their requirements for a limiting soil nutrient and light. Species *A* has the lowest requirement for the nutrient and the highest requirement for light at the soil surface. It has a short, prostrate growth form, a high root:shoot ratio, little stem, and low tissue nutrient concentrations. It dominates nutrient-poor, high light habitats, such as those of early succession. Species *E*, which is the superior competitor for high-nutrient but low light habitats, has the lowest requirement for light but the highest for the nutrient. It is a tall, erect species with a low root:shoot ratio and great allocation to stem. Species *B*, *C*, and *D* are intermediate in their requirements for nutrients and light. If plant species are differentiated in their competitive abilities for a limiting soil nutrient and light, as assumed for these five species, then each species will reach its peak abundance at a different point along a soil-nutrient:light gradient (Fig. 7.1B). If, in primary succession, competitive interactions tend to approach equilibrium at any given point in time, but soil nutrient levels tend to change slowly through time, then the gradient from habitat 0 to habitat 4 could represent a gradient through successional time. The early successional species, with their superior competitive abilities for the limiting nutrient, would attain dominance early in primary succession when nutrient levels were low. As nutrient levels increased and light availability at the soil surface was reduced, these species would be displaced by a series of species that were increasingly superior competitors for light but increasingly inferior competitors for the limiting soil resource. The successional sequence of dominant plant species is thus predicted to be determined by the processes controlling the supply rates of the limiting soil resources.

Species separation along a resource ratio gradient is predicted to occur whenever species are differentiated in their requirements for two or more limiting resources. Such differentiation in plant requirements for light and a soil resource could result from the root-stem-leaf allocation tradeoff. The resource ratio hypothesis does not imply that only the ratio of limiting resources need be known to understand succession, or that succession need be a directional process or have a fixed endpoint. Its essential feature is that changes in the relative supply rates of limiting resources should lead to changes in the composition of a plant community. Plant consumption, disturbance, and various biogeochemical processes can all play a role in

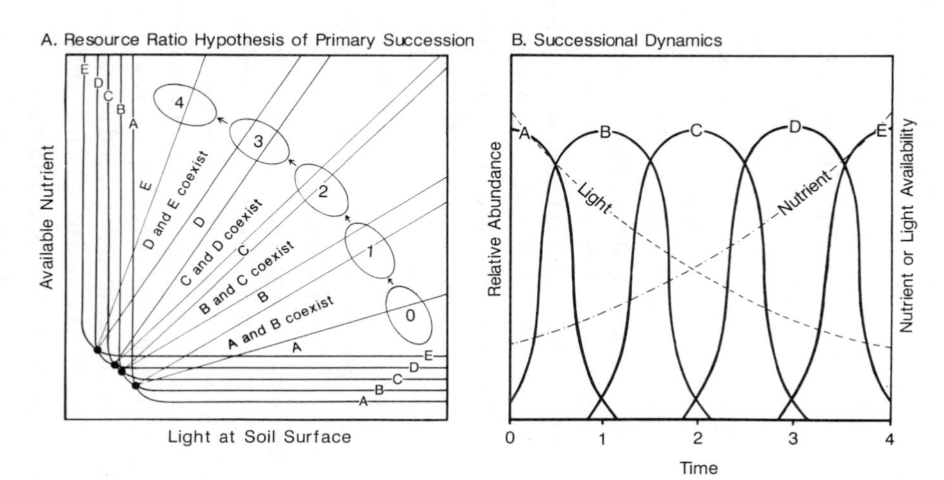

FIGURE 7.1. (A) The curves labeled *A* through *E* are resource-dependent growth isoclines for species *A*–*E*. These species are assumed to be differentiated in their requirements for a limiting soil resource and light, with each consuming these resources in the proportion in which it is equally limited by them. The four dots show stable two-species equilibrium points. The ovals illustrate a primary succession in which a habitat starts with a nutrient-poor soil but high light availability, and then changes into a habitat with a rich soil but low availability of light at the soil surface. (B) Such a trajectory of resource supply rates would cause a successional replacement of species and changes in resource availabilities, as shown. This has been called the "resource ratio hypothesis" of succession.

changing the relative availability of limiting resources. This hypothesis assumes that succession will be a directional and repeatable process only to the extent that the factors controlling the supply of the limiting resources are directional and repeatable. If, for instance, resource supply rates tended to "loop back," succession would not be progressive and directional but "retrogressive." For instance, the classical primary succession of Glacier Bay describes vegetation change during the first few hundred years. However, the invasion of forested stands by *Sphagnum* and/or the formation of an iron hardpan causes soils to acidify and makes them less aerobic. Both of these effects are likely to cause lower rates of nitrogen mineralization, which leads to reduced plant biomass, increased light penetration, and the replacement of a closed-canopy forest by a more open, shorter, muskeg-like vegetation. A qualitatively similar retrogression occurs in the sand dunes of eastern Australia (Walker et al. 1981). The initial succession after sand dune formation is much like the classical cases described for North America, with a closed-canopy forest coming to dominate the dunes. However, slow leaching of phosphorus and calcium coupled with extremely low rates of atmospheric input causes these nutrients to be significantly depleted in dunes during the period from about 100,000 to 400,000 yr after dune formation. As levels of these nutrients decline, plant biomass declines, light at the soil surface increases, and dwarf trees and shrubs replace what had been a closed-canopy forest with 30 m tall trees. Because the resource ratio hypothesis does not assume directional and repeatable changes in resource availabilities, it separates the study of succession, which I consider to be plant dynamics after a disturbance, from the confines of the concept of climax (Clements 1916).

In any terrestrial habitat in which a soil resource is in short supply, as it is during the early stages of primary successions, increased availability of the soil resource would lead to increased plant biomass and thus to decreased penetration of light to the soil surface. This would create a temporal soil-resource:light gradient, much like the gradient illustrated in Figure 7.1, along which species should become separated.

It is critically important to realize that this hypothesis does not state that succession occurs as light becomes a more limiting resource than some nutrient. Rather, it states that such a temporal gradient will sequentially favor species that are superior competitors at particular soil-resource: light ratios. If two species coexist in any habitat, they do so, according to this theory, because one is relatively more limited by one resource and the other is relatively more limited by another resource. A species that is a superior competitor at a particular point along a soil-resource:light gradient attains its superiority because it has a physiology, morphology, and life history that causes it to be almost equally limited by *both* resources at that point. Soil algae are dominant on extremely nutrient-poor soils because they are superior competitors for very low nutrient but high light conditions. They coexist with a short-statured annual on slightly richer soils because the alga, a superior competitor for the nutrient, is light limited, whereas the annual, a superior competitor for light compared to the alga because of its height, is nutrient limited. The annual displaces the alga when both become light limited on an even richer soil. However, on such a richer soil, the annual could coexist with an even taller perennial because each could be limited by a different resource, the annual by light and the perennial by the nutrient. Throughout the full range of nutrient supply rates, both light and the soil resource are predicted to remain limiting to the dominant species. Even though light availability at the soil surface decreases and nutrient availability increases, the resource ratio hypothesis predicts that some of the dominant plants will be limited by the nutrient on all but the richest soils. This occurs because these species are only able to be superior competitors on these high nutrient but low light soils by giving up their competitive ability for a limiting soil resource in order to gain competitive ability for light. This prediction is consistent with the observation that the dominant tree species in equilibrial forests on some of the world's richest soils are still nutrient limited (e.g., Miller and Miller 1976; Albrektson et al. 1977; Waring et al. 1978; Brix 1983). Similarly, Vitousek's (1982) review of nutrient cycling and nutrient use efficiency in a wide variety of ecosystems indicated that a soil nutrient, most often nitrogen, is limiting even in mature stands on rich soils. It is further supported by the results of fertilization experiments in natural vegetation. For instance, the Cedar Creek fertilization experiments (Tilman 1983, 1984, 1987a) showed that nitrogen was the only consistently limiting soil resource in five different fields ranging from an extremely nutrient-poor early successional field to much more nutrient-rich native oak savanna. However, only about half of the common species increased in response to nitrogen addition in any of the fields. The other half tended to decrease following nitrogen addition. Such a pattern is consistent with the existing species in a given field being near the boundary at which each is almost equally limited by nitrogen and another resource, most likely light.

EDAPHIC FACTORS AND COMMUNITY COMPOSITION

The theory summarized in Figure 7.1 predicts that, at equilibrium, the composition of a plant community will be greatly influenced by the rate of supply of the limiting soil resource. Such edaphic (i.e., soil-caused) variation in vegetation composition has been repeatedly observed. Whittaker (1953, p. 44) stated that "references to edaphic factors occur throughout the literature, such effects apparently being observed in every vegetational area studied with sufficient intensity." As Jenny (1980) noted, on broad geographic scales the major factors determining vegetational composition are climate and soils. On smaller scales, climate is much less variable, but soils can be extremely heterogeneous. Soil heterogeneity is caused by differences in parent material, its slope, its aspect, and its disturbance history. Chance events of deposition as well as the more repeatable process of erosional separation and deposition all contribute to the heterogeneity of soils.

The work of Pastor et al. (1982) is an excellent example of the correspondence between the local rate of nitrogen mineralization and the species composition of the vegetation that comes to dominate a site. It demonstrated a clear correlation between the initial parent material (specifically its clay content), the nitrogen mineralization rate of the soil that developed on the parent material, and its eventual vegetation. Zak, Pregitzer, and Host (1986) performed a study in Michigan similar to that of Pastor et al.'s (1984) on Blackhawk Island in Wisconsin, but sampled 12 different stands that were spread over a much broader landscape range than those on Blackhawk Island. Four of these stands were composed of a mixture of black oak and white oak with a *Vaccinium*-dominated understory. Four stands contained a mixture of sugar maple and red oak with a *Maianthemum* understory. The other four stands were dominated by sugar maple and basswood with an *Osmorhiza* understory. These stands, in this order, represented a gradient from a relatively open-canopy forest to a more closed-canopy forest, to a forest with a highly closed canopy. They collected soils from each stand and incubated the soils at constant temperature and moisture to determine the potential rates of nitrogen mineralization for each soil type. They found that the soils of the stands dominated by black oak and white oak had significantly lower rates of nitrogen mineralization than the other two communities, and that the sugar maple and red oak stands had significantly lower rates of nitrogen mineralization than the sugar maple and basswood stands (Fig. 7.2).

Thus, in stands separated by at least 6 km, the rate of nitrogen mineralization of the soil was a good predictor of its vegetational composition (Zak et al. 1986). The one exception was a sugar maple stand that had the lowest rate of nitrogen mineralization of all 12 stands (Fig. 7.2). This stand was resampled to determine if the initial results could in some way have been caused by an unrepresentative series of soil samples, and was found to have a low rate of mineralization the second time, too (Zak, personal communication). This stand, though, has unusually low plant biomass and standing crop, and has high light penetration to the soil surface. Zak believes that its topsoil was lost to erosion during clearcutting but that sugar maple and basswood reestablished themselves because of their high seed source and the low seed source for black and white oak. If this is so, this stand should be invasible by oaks, which should be able eventually to reduce the density of maples and basswood. Across all these stands, including the unusual maple-basswood stand, the stands with higher rates of nitrogen mineralization had a more closed canopy and lower light penetration to the soil surface, just as Pastor (Fig. 5.6) found for Blackhawk Island.

These results are amazingly consistent with observations that Olson (1958) made in the sand dunes of Indiana along southern Lake Michigan. Olson reported that the old-growth, late successional vegetation of this region depended on the nitrogen content of its soil. Areas that developed a lower-nitrogen soil were dominated by black oak with a *Vaccinium* understory, whereas areas that developed a more nitrogen-rich soil were dominated by basswood-sugar maple forest. Robertson and Vitousek (1981) and Robertson (1982) performed studies of potential nitrogen mineralization on these sand dune soils that demonstrated that the soils with lower levels of total soil

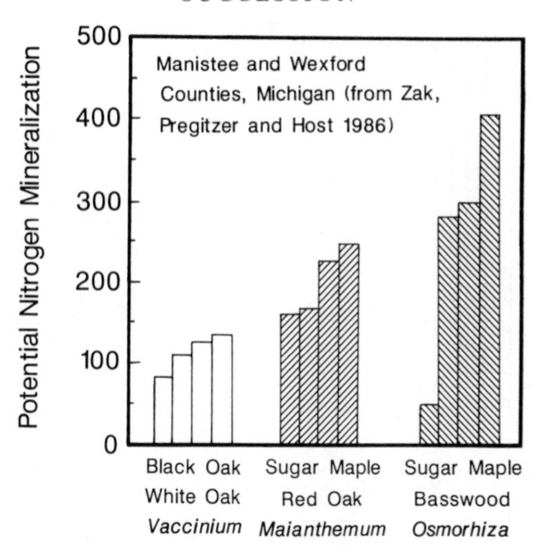

FIGURE 7.2. Rates of nitrogen mineralization measured under laboratory incubation conditions for soil samples collected in 12 different forest stands in northern Michigan. Redrawn from Zak et al. (1986).

nitrogen also had lower rates of mineralization, further bolstering the similarities between the work by Zak et al. (1986) and that performed in the Indiana sand dunes. Thus, across distances of hundreds of kilometers, the composition of the vegetation that eventually comes to dominate a particular site seems to be determined in a qualitatively similar manner by the rate of supply of a major limiting soil resource. The separation of these overstory tree species with respect to soil nitrogen and the degree of light penetration to the soil surface suggests that these species differ in their requirements for nitrogen and light. The expected differences among these species can be illustrated using resource-dependent growth isoclines (Fig. 7.3). The relations shown in Figure 7.3 are a hypothesis that could be tested by determining the level to which available soil nitrogen and light are reduced in monocultures of these

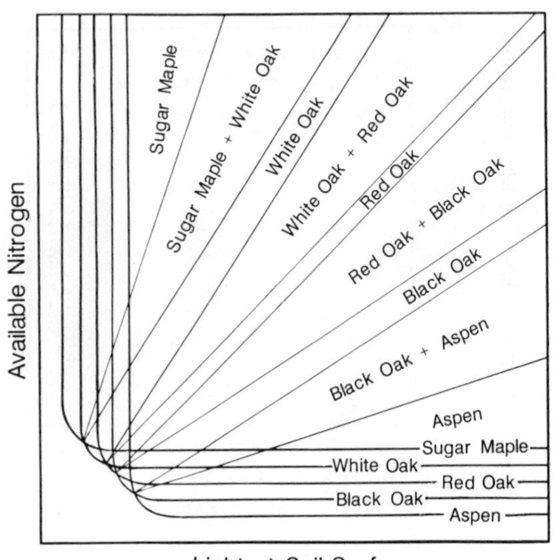

FIGURE 7.3. Hypothesized resource-dependent growth isoclines for 5 common trees of central North America.

species grown to equilibrium on different soil types. These data are not as difficult to obtain as it might seem initially. Several of these species are important in forestry, and foresters have established many monospecific stands.

Although the work of Pastor et al. (1984), Olson (1958), Robertson and Vitousek (1981), Robertson (1982), and Zak et al. (1986) shows a strikingly similar relationship between vegetational composition and nitrogen supply rate, such correlational studies cannot demonstrate a cause-and-effect relationship. Their results could be, and have been, interpreted as indicative of vegetational effects on nitrogen supply rather than nitrogen supply effects on vegetation. As already discussed, the reason for such alternative interpretations is clear: soils affect plants *and* plants affect soils. The best way to overcome this problem is through direct experimental manipulation of limiting resources, including light. However, such experiments must continue for many years to be of much use in understanding forest vegetation.

Alternatively, some insights may be gained by looking at the relationship between vegetation and the parent material upon which a soil is formed. I have already mentioned that on Blackhawk Island there was a good correspondence between subsurface (and presumably initial) clay content and eventual soil and vegetation. Jenny (1980) asserted that the original parent material was one of the major soil-forming factors. Thus, there is often a strong correlation between the initial substrate of a locality and the composition of the vegetation that eventually comes to dominate that locality. Clearly, parent material is not the only factor determining vegetational composition. Disturbance history, fire frequency, local herbivore densities, and climate all influence the vegetation of an area. However, once such factors are controlled, there is often a strong dependence of vegetation on parent material.

For example, the soils within Wisconsin that were formed on deep (> 1 m), acid, glacial outwash sand were mainly dominated by pine barrens before settlement (Fig. 7.4; Hole 1976). The other vegetation types common on these thick sands were oak savanna, prairie, and pine forest (Hole 1976, p. 73). Hole's (1976) maps of soils and vegetation of Wisconsin show many other striking similarities between parent materials and vegetation, many of which reinforce the patterns reported by Pastor et al., Olson, Zak et al., and Robertson and Vitousek. Whitney's (1986) study of original vegetation and soils of several Michigan counties also supports their work. Similarly, soil and vegetation maps of Minnesota show a strong correspondence between the original substrate and the eventual dominant vegetation within an area. As in Wisconsin, the large sand plains of Minnesota are dominated by pine barrens, oak savanna, or prairie. Glacial till in both localities most often supports a closed-canopy vegetation such as oak forest or sugar maple-basswood forest.

Lindsey (1961, p. 434) found that patterns in undisturbed vegetation of northern Indiana "definitely tend to

FIGURE 7.4. (A) Location of deep sandy soils underlain by outwash sands in Wisconsin. (B) Location of pine barrens in the original vegetation of Wisconsin. Redrawn from Hole (1976).

follow substrates produced by different modes of glacial action, even though I minimized edaphic differences by mapping forest types as they occurred on only well-drained median portions of the terrain mosaic." Similarly, Rabinovitch-Vin (1979, 1983) found, for the Upper Galilee of Israel, that "different climax plant communities grow in the same altitudinal belt under similar climatic conditions." Rabinovitch-Vin attributed these differences in vegetational composition to the type of parent material on which a soil formed. There have been a wide variety of studies demonstrating a dependence of vegetation on soils, whether the study has been performed over distances of meters, kilometers, or involved comparisons of different continents (references above; Beard 1944 and 1955; Box 1961; Snaydon 1962; Pigott and Taylor 1964; Beals and Cope 1964; Zedler and Zedler 1969; Hanawalt and Whittaker 1976; Cody and Mooney 1978; Beard 1983).

In total, these studies demonstrate that soil characteristics, including original parent material, may be a major factor controlling the abundances of species during primary succession. They support the view that soil resources and light are inversely correlated both in space and in time, and that plants are adapted for particular points along soil-resource:light gradients. Such evidence runs counter to Grime's (1979) claim that a species that is a "good competitor" is a good competitor for all resources. Further, these studies suggest that primary successions will reach a stable endpoint only if the various factors influencing soil resource supply rates cause resource supply rates to equilibrate.

RESOURCES AND SECONDARY SUCCESSION

The dynamics of primary succession, as discussed above, are strongly influenced by soil dynamics, with soil formation occurring at a rate similar to the rate of competitive displacement. This slow rate of soil formation reduces the importance of a variety of factors that can influence secondary succession on a rich soil. Some secondary successions occur on soils that are sufficiently nutrient rich that nutrients do not accumulate during succession, but may actually be lost (Aarsen and Turkington 1985). Other secondary successions occur on nutrient-poor soils that have a period of nutrient accumulation much like that of primary successions (e.g., Billings 1938; Odum 1960; Rice et al. 1960; Inouye et al. 1987a). Despite such differences in their nutrient status, there are many qualitative similarities among primary successions, secondary successions on poor soils, and secondary successions on rich soils. The initial dominants of secondary successions on both rich and poor soils tend to be herbaceous, fast-growing, short-lived species that have a short stature at maturity, whereas the later dominants are often woody, slow-growing, long-lived, tall perennials. Primary successions also have a progression from short, often herbaceous species, to tall, woody species.

Such similarities, though, contradict the role of a soil-resource:light gradient as a general explanation of all successions. Secondary succession on a rich soil starts with high availability of both nutrients and light. The condition of high light availability but low nutrient availability for which early successional species, such as species *A* and *B* of Figure 7.1, are superior competitors, at equilibrium, never occurs. However, as discussed in Chapter 6, the dynamics of competitive displacement within a given habitat (i.e., fixed resource supply and loss rates) are strongly influenced by the maximal growth rates of the competing species. Within the constraints imposed by the availabilities of the limiting resources, the initial dominants in a newly disturbed, low plant density habitat are species with rapid rates of growth. These are displaced by a series of species that have lower maximal growth rates but are increasingly superior competitors for the conditions of that habitat (Figs. 6.2–6.6).

Thus, if species differ in their maximal growth rates, a succession of species are expected to attain dominance during the process of competitive displacement. This suggests that the successional sequence observed during secondary succession on rich soils may be just the transient dynamics of competitive displacement. If this is so, what would be the expected similarities and differences between secondary successions on rich soils and those on poor soils, assuming that other processes, such as colonization and herbivory, were similar?

Transient Dynamics

Two different processes are at work determining which species are abundant at a particular time during the dynamics of secondary succession. First, there are transient dynamics that are influenced by maximal growth rates, allocation patterns, and the relatively high initial availabilities of all resources when population densities are low at the start of succession. Second, there is the process of long-term soil change, such as the long-term accumulation of nitrogen observed during secondary successions on poor soils. All cases of secondary succession, whether on poor or rich soils, are expected to have an initial period that is mainly controlled by transient dynamics. For instance, consider a case in which 100 different species compete on an initially bare, nutrient-poor soil (TN = 100 initially). Over a period of 100 years, the nutrient level of the soil gradually increases to 1000. The dynamics of the first 10 years of this secondary succession are mainly transient dynamics that are most strongly influenced by differences in the maximal growth rates of the species (Fig. 7.5A). After this time, the observed dynamics are more strongly influenced by the slow accumulation of nutrients, with the composition at any given time being determined by the recent nutrient richness of the soil. The trajectory of traits of the species dominant at different times during this secondary succession on a poor soil is from species with high maximal growth rates to those with low maximal growth rates. It is also a trajectory first from short species with high leaf allocation but low allocation to stem and moderate allocation to roots, to short species with higher root but lower leaf allocation, and then to taller species with increasingly greater stem allocation but increasingly lower root allocation (Fig. 7.5A). The initially dominant morphologies—the leafy, fastgrowing species—are never a part of the equilibrium community expected on any of these soil types for a habitat with this loss rate. They are truly transients that attain a period of dominance because their morphologies allow them to grow rapidly and temporarily competitively suppress other species for the particular nutrient and light levels that occur on the newly abandoned soil. The prediction that there can be a period of transient dominance on poor soils differs from Tilman (1985), in which I stated that transient dynamics should only be important during secondary succession on a rich soil. The model upon which that conclusion was based, however, was not morphologically explicit, but assumed that "early successional" species were limited to species of nutrient-poor habitats.

The trajectory of dominant species during secondary succession on a more nutrient-rich soil (TN of 200 initially and of 1000 at 100 years) also has a long period during which transient dynamics are of greater importance (Fig. 7.5B). On an ever richer soil (TN of 1000 both initially and after 100 years), the full successional sequence is determined by transient dynamics (Fig. 7.5C). Because all of these are constrained to having the same final soil, and because all have the same background loss rate, the final dominant species in all these cases of secondary succession are the same. Moreover, the initial dominants for all cases of secondary succession share certain features. All are short, leafy, fast-growing species. Thus, this model predicts some qualitative similarities between secondary successions

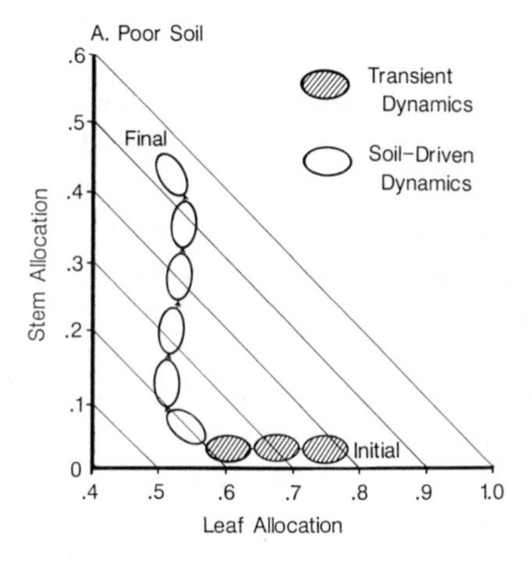

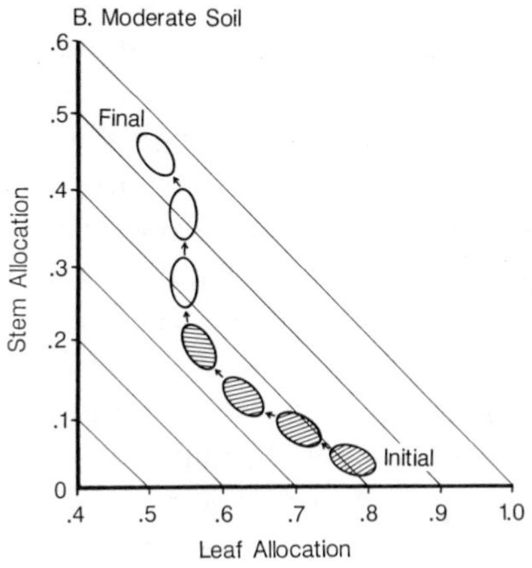

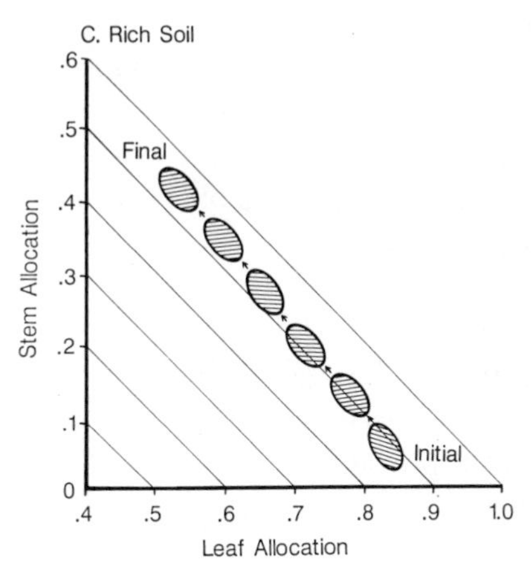

FIGURE 7.5. The roles of transient dynamics and long-term nutrient accumulation (termed soil-driven dynamics) in controlling succession on soils of different initial nutrient richness. In all cases, the soil was assumed to attain a TN of 1000 after 100 years. The shaded ellipses show the sequence of dominant species (morphologies) that was caused by transient dynamics. The open ellipses show the sequence caused by nutrient accumulation during the 100 years. (A) The period of transient dominance lasted about 10 yr on the poor soil (TN = 100), followed by a 90 yr succession that was driven by nutrient accumulation in the soil. (B) The transient dynamics lasted about 7 yr on the moderate soil (TN = 200 initially). (C) On the rich soil (TN = 1000 initially), the full dynamic sequence is purely transient dynamics. These transient dynamics lasted about 6 years. Note that this soil started with TN = 1000, and thus no accumulation occurred. These simulations assumed that all soils would eventually attain TN = 1000, and thus have the same equilibrial community. Clearly, that need not be the case in most successions. This was assumed purely to allow easy comparison of these three cases.

that start on soils of different initial nutrient status. However, the actual sequence of dominant species is predicted to differ, with species specialized on nutrient-poor, low loss rate habitats attaining dominance during secondary succession on poor soils, but never attaining dominance during secondary succession on rich soils (compare parts A, B and C of Fig. 7.5). Conversely, species that would be equilibrial dominants of nutrient-rich, high loss rate habitats are predicted to attain a period of transient dominance during secondary succession on rich soils but not on poor soils.

These results suggest that there are many different traits that could lead a species to be an "early successional" species. During primary succession in habitats with moderate loss rates, the initial dominants are predicted to be species that have low maximal growth rates because their high allocation to structures (roots and mycorrhizal fungi) or processes (nitrogen fixation or highly efficient uptake systems) for nutrient acquisition gives them lower allocation to leaves. Faster growing species that have higher requirements for nutrients are simply unable to grow, and would not attain a period of transient dominance. In contrast, during secondary successions on a rich soil, the initial dominants are species that attain a period of transient dominance because of their rapid growth rates. Further, all else being equal, species with greater colonization abilities would be able to attain a period of transient dominance during early succession. There seems to be little information conveyed by classifying a species as "early successional" or "late successional" without also stating the habitat conditions for which it becomes dominant.

Weed Evolution

The theory presented in this book suggests that two of the major axes along which plant species have become differentiated are soil resource supply rate gradients and loss rate gradients. The physiology and morphology of an individual plant constrains it to being a superior competitor at a particular nutrient supply rate and a particular loss rate. The advent of agriculture and other anthropogenic disturbances has led to an explosion of high loss rate habitats. During the period of expanding agriculture from Neolithic times to the present, it is likely that a variety of species evolved traits that made them better competitors for high loss rate habitats. Many of these would have been derived from species that had short life cycles, were non-woody, and had relatively low allocation to stems. The most likely ancestors for current weeds are species that were superior competitors in open, relatively low-nutrient soils in a seasonal environment. These would be the species dominating sand plains, gravel bars, rock outcrops, eskers, limestone pavements, and other "waste lands." An admittedly informal reading of Gleason and Cronquist's (1963) flora of North America and of Clapham, Tutin, and Warburg's (1962) treatment of the British flora indicates that these are just the types of natural habitats in which species that have become agricultural weeds live.

What processes may have allowed species that are now dominant in early successional fields to persist within a geographic area before the advent of massive anthropogenic disturbances? The most common explanation is that such species are maintained because of a pioneer, fugitive lifestyle, i.e., that their superior colonization rates and rapid growth rates allowed them to exploit small-scale disturbances in primeval habitats. Marks (1983) addressed this question by asking where, in the primeval landscape of eastern North America, did the native plants that are now early dominants of old fields persist before agriculture. He used habitat descriptions from old floras, plant compositions of current forest openings in otherwise native forest stands, and numerous observations on vegetation stands that had been essentially undisturbed since colonial settlement of North America as the basis for his work. He found that the native plants that are today associated with newly abandoned agricultural fields occurred "primarily in persistent open habitats rather than in temporary forest openings" (Marks 1983, p. 225), thus suggesting that these species that today function as early successional species probably did not evolve as fugitives specialized on locally disturbed sites within forests. Furthermore, he said that "most native field plants have evolved in marginal habitats that are permanently open on the time scale of centuries or millennia" (p. 225). These persistent habitats were areas such as limestone outcrops, gravel bars, sand plains, talus slopes, rock crevices, ridgetops, rocky and sterile soils, river islands, and steep, eroded stream banks. Thus, all of them are high light habitats. Many of them are high light habitats because of poor soils, but some of them, such as the eroded stream banks and the river islands, may also have high loss rates.

The early successional species that Marks considered, such as little bluestem, horseweed, several goldenrods and several woody shrubs, are all plants of short stature at maturity that tend to have greater maximal growth rates than the trees, such as sugar maple and beech, that dominate rich soils in eastern North America. Thus, Marks' observations support the view that many species that are common during early secondary succession did not evolve their life history characteristics to allow them to have a fugitive existence in disturbed patches in the "climax" vegetation of areas with rich soils, but rather were adapted to permanently open habitats with poor soils or, possibly, high loss rates.

OTHER VIEWS OF SUCCESSION

I have emphasized the role of resource competition in succession, but not because it is necessarily of overriding importance. There are several other processes besides resource competition and the transient dynamics of competitive displacement that could also be major determinants of successional patterns. One of these is colonization, which is the first step in the successional process. A species cannot become a dominant if it never reaches a habitat. Nor can it dominate a habitat if it cannot survive and grow there. However, even if we only consider that pool of potential colonists that are capable of dominating a habitat in the absence of interspecific competition, colonization, by itself, cannot explain succession. I know of no studies of succession that have shown that the first species to grow on a site can hold that site indefinitely. A viable "colonization hypothesis" of succession requires that there be a tradeoff between colonization ability and some other aspect of a plant's ability to survive and grow. For instance, a colonization-competition hypothesis would assume that species that are better colonists are poorer competitors, as is suggested by the work of Werner and Platt (1976). Succession would then occur as a bare site was populated first by rapid colonists, which were later displaced by a series of species that were poorer colonists but superior competitors. The site would ultimately be dominated by the species that was the poorest colonist but the best competitor. A colonization-herbivory hypothesis would assume that species that were good colonists were more sensitive to herbivory, and would be displaced by later colonists that were more resistant to herbivory. Colonization is likely to be of increasingly great importance on nutrient-rich sites because plant growth is rapid on such sites. When plant growth is rapid, initial colonists can attain a period of dominance before other species that are better suited to the conditions arrive. However, all primary successions and some secondary successions start with a nutrient-poor soil. Because plant growth is slow on poor soils, there is a longer period during which colonization may occur before any species has the opportunity actively to displace other species. At Glacier Bay Alaska, for instance, many of the later dominants of succession were among the first species to colonize newly exposed glacial till (Lawrence 1979). This rapid colonization, though, did not allow them to be the first dominants because they were incapable of growing on these extremely nutrient-poor soils (Lawrence 1958; Crocker and Major 1955). Thus colonization is likely to be a less important process during primary succession and secondary succession on poor soils than during secondary successions on rich soils.

Another possible explanation for the pattern of succession could be called the competition-herbivory hypothesis. This hypothesis assumes that there is a tradeoff in the competitive abilities of species versus their susceptibility to herbivory, with superior competitors being more susceptible to herbivory. If this were so, any directional change in the herbivory rate would lead to a directional change in the composition of the vegetation. For instance, if rates of herbivory were low during early succession and increased through time, there should be a successional sequence from species that were good competitors but susceptible to herbivory to species that were poor competitors but resistant to herbivory. If the rate of herbivory decreased during succession, the plant successional sequence would be from poor competitors that were resistant to herbivory to good competitors that were not resistant to herbivory.

SUMMARY

Colonization, competition, transient dynamics, herbivory, and other processes can all influence successional dynamics. It is likely that the relative importance of these processes changes from habitat to habitat. As such, there may be little reason to seek a single universal explanation for all successions. However, succession is interesting because, despite the differences among habitats, there is a feature shared by almost all successions: the initial dominants are herbaceous plants of short stature that are replaced by a sequence of increasingly taller, often woody, species. This similarity is probably caused by all successions starting as high light habitats, and by shorter plants being superior competitors in high light habitats. The theory developed in this chapter predicts, despite this similarity, that the actual sequence of species that are dominant during succession should differ for primary successions, secondary successions on poor soils, and secondary successions on rich soils. This occurs because each of these successions has a different trajectory of nutrient and light availabilities and because the morphology and physiology of each species constrain it to being a superior competitor in a habitat with a particular nutrient and light availability. Moreover, long-term, slow changes in resource availabilities are likely to be a more important cause of primary succession, and the transient dynamics of competitive displacement are likely to be of greater importance on substrates that are initially more nutrient rich. On very rich soils, the entire successional sequence could be just the transient dynamics of competitive displacement. On poorer soils, the initial dynamics may be mainly transient dynamics, and the later dynamics may be more strongly influenced by a period of nutrient accumulation in the soil. These forces shaping the dynamics of succession, though, interact with other processes, most notably colonization and herbivory, in any actual succession. The source pool of species that could potentially colonize a habitat is also a critical determinant of its successional dynamics. Although succession is a multicausal process, the constraints that allocation patterns place on plant life histories and morphologies, and the effects of these constraints on the dynamics of competition may be a major cause of the qualitative similarities of successions worldwide. These constraints lead to the prediction that all successions should progress, at least initially, from short to tall species. Further, they predict that secondary successions should progress, initially, from rapidly growing species to more slowly growing species.

CHAPTER EIGHT

Secondary Succession on a Minnesota Sandplain

In 1982, I initiated a series of observational and experimental studies on the controls of the composition, diversity, and dynamics of the old fields of Cedar Creek Natural History Area. In this chapter I will use the results of these studies to date to evaluate many of the ideas developed in this book. Although some of the material presented in this chapter has been separately published (Tilman 1983, 1984, 1986a, 1986b, 1987a; Inouye et al. 1987a, 1987b), much is presented here for the first time. I should note that these experiments include more than 1000 different plots (most of them 4 × 4 m) and a wide range of manipulations, only some of which can be discussed here.

Cedar Creek occurs on a large, deep outwash sandplain that was deposited by meandering glacial streams about 13,000 years ago (Fig. 8.1). At the time of European settlement in the mid 1800's, the vegetation of the upland areas of this sandplain was a mixture of prairie opening, oak savanna, and oak forest (Fig. 8.1). The soils that formed on these upland, excessively well-drained sands had low organic content and were nitrogen poor (Grigal et al. 1974; Tilman 1983, 1984, 1987a). Nearby soils that formed on glacial till or the prairie soils to the south and west that formed on wind-deposited loess, had from 2 to 5 times higher organic matter and total nitrogen content. Between about 1890 and 1910, many of the upland areas at Cedar Creek were clearcut and farmed. From 1930 to the present time, many of these fields have been abandoned from agriculture, and have undergone secondary succession, but in the absence of fire. Before settlement, this region tended to have fires once every one to four years (Grimm 1984). With settlement came fire suppression, which has led to the spread of oak savanna into former prairie openings and the formation of a more closed-canopy oak forest in areas that had been oak savanna (White 1983).

THE SUCCESSIONAL SEQUENCE

The first dominants of newly abandoned old fields are such annual plants as *Ambrosia artemisiifolia* (ragweed), *Polygonum convolvulus* (bindweed), *Setaria glauca* (foxtail), *Erigeron canadensis* (horseweed) and *Chenopodium album* (lamb's quarters). These annuals will often cover 10 to 40% of the soil surface. Perennials, especially such short-lived perennials as *Agrostis scabra* (bent grass), may cover an additional 5 to 10%. The remainder of the soil surface is bare. During the first 60 years of succession, annual plant cover decreases, perennial plant cover increases, and the proportion of the soil surface that is not covered by plants decreases (Fig. 8.2; Inouye, Huntly, Tilman, Tester, Stillwell, and Zinnell 1987). The perennials that become increasingly dominant are native prairie grasses such as *Schizachyrium scoparium* (little bluestem), *Andropogon gerardi* (big bluestem), *Sorghastrum nutans* (Indian grass), and native sedges such as *Carex muhlenbergii*, as well as the introduced grass *Poa pratensis* (Kentucky bluegrass). Although forbs as a group become less abundant during succession, prairie forbs are increasingly abundant in older fields. Non-vascular plants, mainly lichens, are rare in both newly abandoned fields and in old fields, and reach their peak abundance in 20- to 30-year-old fields (Fig. 8.2). Woody plants, including vines such as *Rubus* sp. (blackberry), small bushes such as *Rosa arkansana*, and trees such as *Quercus macrocarpa* (bur oak), *Quercus borealis* (red oak), and *Pinus strobus* (white pine) are essentially absent from fields younger than about

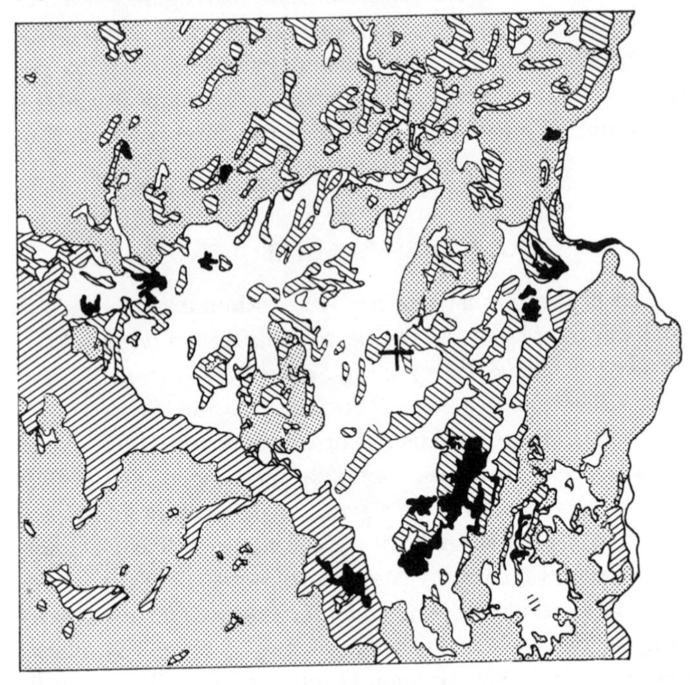

FIGURE 8.1 (A) Soils of east-central Minnesota. The area shown is approximately 85 km x 85 km. The large, light-colored area is the Anoka sandplain, on which Cedar Creek Natural History Area is located. The location of Cedar Creek Natural History Area is indicated by the "+." Map based on the Stillwater sheet of the Minnesota Soil Atlas, Misc. Report 171-1980, Agricultural Experiment Station, University of Minnesota, St. Paul, MN.

B. Vegetation of East Central Minnesota

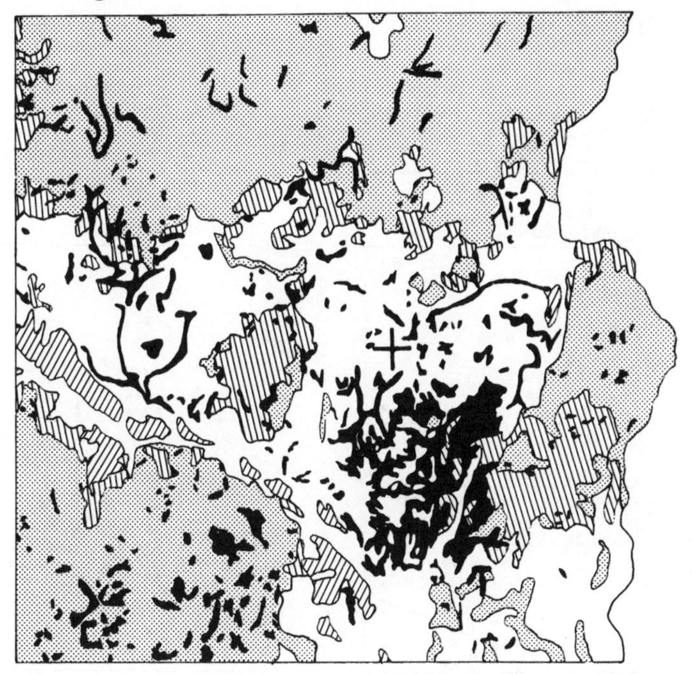

Wet Prairies (marsh grasses and reeds)

Oak Openings (scattered bur oaks and prairie)

Prairie

Aspen-Oak (aspen and scattered oaks)

Big Woods (maple, oak, elm, basswood and ash)

(B) Original (presettlement) vegetation of this region, based on the map of "Original Vegetation of Minnesota" by F. J. Marchner (1930), republished by the North Central Forest Experiment Station, St. Paul, MN. Both maps redrawn to the same scale.

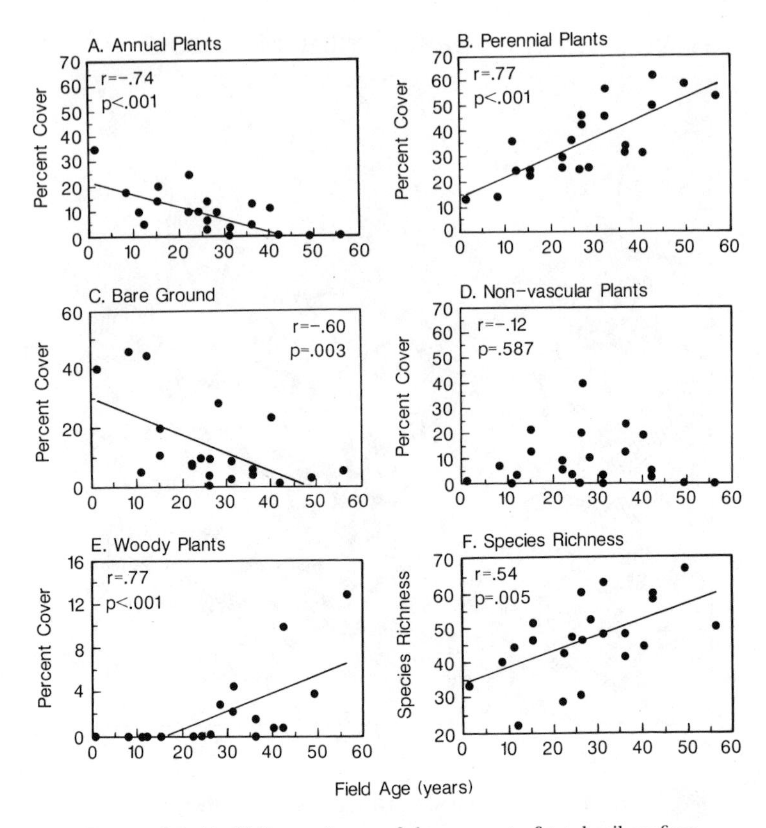

FIGURE 8.2. (A–E) Dependence of the percent of total soil surface covered by annuals (A), perennials (B), bare mineral soil (C), non-vascular plants (D), and woody plants (E) in relation to the age of 22 Cedar Creek successional fields surveyed in 1983. (F) Total number of plant species observed in each old field in relation to field age. Redrawn from Inouye et al. (1987a).

15 years, and increase slowly in cover through time (Fig. 8.2). Even in 60-year-old-fields, woody plant cover is, at most, 13%, and much of this comes from vines and shrubs, not trees. The low cover of woody plants in these older fields is surprising because most of the fields are rather small (less than 5 or 10 ha) and are surrounded by woodlands that have never been clearcut. As these changes occur in the species composition of these old fields, the total number of species per field increases from an average of about 30 in newly abandoned fields to about 60 species in the oldest fields (Fig. 8.2).

What causes such changes in the species composition and diversity of these Cedar Creek old fields? Clearly, many processes may be involved. Part of the observed successional dynamics may be transient dynamics caused by differences in colonization rates and/or growth rates. The plant species may differ in their competitive abilities, and their competitive abilities may change through time as habitat conditions change. Further, herbivore densities may change during succession, and thus influence the pattern of succession. A simple explanation for the initial dominance by annuals is that they were already present in high densities when the fields were abandoned from agriculture. The frequent tilling associated with farming selects for plants with rapid growth rates and short life histories. Thus, at the time of abandonment, seeds of annuals were already quite abundant in the soil. I have often observed densities of annual plant seedlings greater than 1000/m^2 in newly abandoned fields. Unlike many other successions, though, annuals remain a major component of the vegetation for decades. Annuals accounted for over 40% of the total plant cover (which equals 100% − % litter cover − % bare ground) in a 20-year-old field, and over 20% of the total plant cover in a 40-year-old field (Fig. 8.2). These fields had not experienced major disturbances or high loss rates that could account for the high proportionate cover of annuals.

Thus, although the initial dominance by annuals may be explained purely as an artifact of their initially great abundance and their greater growth rates, their persistence as a major component of the vegetation of these fields for up to 40 years cannot be so explained. During those 40 years, many other species invaded these fields and became established, and yet did not displace annuals. Not only were annuals not displaced, but total plant cover (i.e., % of ground covered by plants) in these fields had only increased to about 60% after 60 years of succession. About 40% of the soil surface was bare or covered only by a thin layer of litter, even in the oldest fields. Although the rarity of woody plants during the first 5 or 10 years of succession could possibly be explained as a transient effect caused by slow dispersal rates or slow growth rates, this could not explain the failure of woody plants to dominate the oldest fields. Oak seedlings are common in Cedar Creek old fields, and yet saplings are rare. White (1983) found that most of the bur oak on Cedar Creek savanna had attained their size at maturity within 40 years. Thus, though colonization rates and differences in growth rates may account for some of the initial patterns of succession at Cedar Creek, the rate of successional change there is too slow to have this be the only explanation for the patterns we see. In our analyses of the data collected in our survey of 22 Cedar Creek old fields, we found that both field age and the nitrogen status of the soil were important correlates of successional patterns. We concluded that "time, acting through differences in dispersal and colonization rates, may thus play an important role in determining which species are present in particular fields, while resource-based interactions determine which species are found in particular places within fields. Thus, this work suggests that the pattern of secondary succession on a Minnesota sand plain results both from dispersal and other time-dependent processes and from changes in the availability of a major limiting soil resource, nitrogen" (Inouye et al. 1987a, p. 25). Neither process, alone, is sufficient to explain the patterns we observed.

It is possible that herbivores could explain some of these patterns. The major herbivores at Cedar Creek are the plains pocket gopher (*Geomys bursarius*), the white-tailed deer, voles (*Microtus pennsylvanicus*), and several species of grasshoppers. Herbivore exclusion experiments performed at Cedar Creek have shown that each of these can have a detectable effect on the above-ground plant biomass (Huntly et al., in prep.). However, even the exclusion of all herbivores for 4 years increased total plant biomass by only about 10%. Although most herbivores have some preferential feeding, herbivore densities were sufficiently low that this had little qualitative effect on the relative abundances of plant species along soil nutrient gradients. Deer exclosure experiments have similarly shown that there can be heavy deer browsing on the margins of these fields, but that this browsing only slightly reduces the annual height increment in woody plants. Although these herbivores do have an impact, it seems unlikely that they are having a major effect on the course of succession. Plains pocket gophers, for instance, create mounds of bare soil as they excavate their feeding tunnels. Gopher mounds are initially dominated by annuals, especially seed-banking annuals (Tilman 1983). However, no more than 1 or 2% of the soil surface of a normal field is so disturbed in any given year. This rate of soil disturbance is too low to explain the long-term persistence of high densities of annuals in these old fields or the rarity of woody perennials.

Another possible explanation for many of the successional patterns is the nutrient status of the soils. Numerous nutrient addition experiments performed in the old fields and in native oak savannah of Cedar Creek have shown that nitrogen is the most important limiting soil resource throughout succession in these areas (Tilman 1983, 1984, 1987a; Fig. 8.3). Nitrogen addition has consistently led to significant increases in above-ground biomass, but the addition of all nutrients except nitrogen has never led to a significant increase compared to controls (Tilman 1987a). In our survey of a chronosequence of Cedar Creek old fields, we sampled both the soils and the vegetation of 2300 quadrats spread over 22 different fields (Inouye et al. 1987a). These soil samples showed that total soil nitrogen increased highly significantly with field age (Fig. 8.4). Newly abandoned fields had, on average, 370 mg/kg of total soil nitrogen, which is about one-third that of the forests that surround the fields. The observed rate of increase in nitrogen through time inferred from this chronosequence is such that it would take a field about 100 years to have its soil nitrogen level return to that typical of adjacent, native, undisturbed oak savanna or oak forest.

As total soil nitrogen levels increased in these old fields, the total cover by vascular plants increased and that by nonvascular plants decreased (Fig. 8.5). Annual plants and introduced species decreased in abundance along this temporal nitrogen gradient, and perennial herbaceous species and woody plants increased in cover (Fig. 8.5). Thus, many of the same variables that were significantly correlated with field age were also correlated with the nitrogen richness of the field. Samples collected in 19 of these fields in 1984 showed that total above-ground biomass increased significantly with total soil nitrogen and that penetration of light to the soil surface was negatively correlated with total above-ground plant biomass (Inouye et al. 1987b). Further, soil samples collected along the same transects in 16 of these fields in 1986 showed that the level of available soil nitrogen (the sum of extractable NH_4 plus extractable NO_3) was higher in the oldest, most nitrogen-rich fields (Fig. 8.6; Tilman 1987b). Moreover, Pastor, Stillwell, and Tilman (1987) found that the rate of *in situ* mineralization of nitrogen in 4 of these fields was positively correlated with total soil nitrogen. Thus, one component of old field succes-

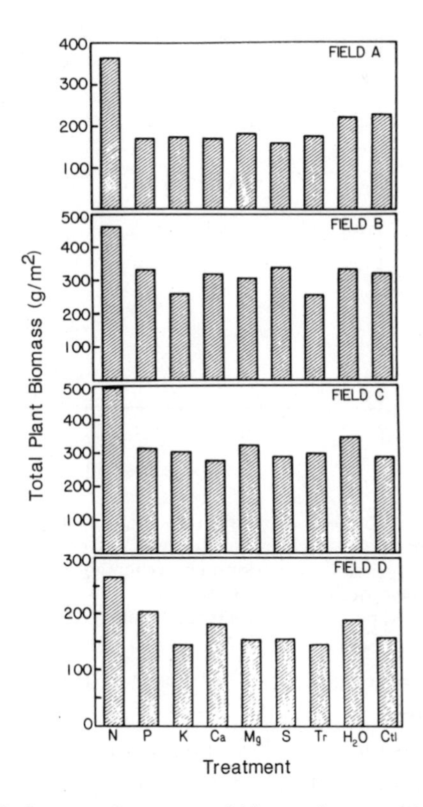

FIGURE 8.3. Average above-ground biomass harvested in each of 9 treatments in a nutrient addition experiment that was replicated in 4 different fields. There were 4 replicates per treatment in each field, giving a total of 144 plots. The fields, described in detail in Tilman (1987a), were: Field *A*, 14 yr old, Field *B*, 25 yr old; Field *C*, 48 yr old; and Field *D*, native oak savanna. Treatments consisted of addition of various nutrients one at a time. N is ammonium nitrate; P is sodium phosphate; K is potassium chloride; Ca is calcium carbonate; Mg is magnesium sulfate; S is sodium sulfate; Tr is a mixture of trace metals (Co, Mn, Mo, Cu, Zn); H_2O is water, added at a rate of 2 cm per week during June, July, and August; Ctl is the unmanipulated control. ANOVA's with contrasts between each treatment and the control showed that, on a year by year basis, or averaging over all 5 harvests from 1982 through 1986, whether on a field by field basis, or blocking for field effects, nitrogen was the only nutrient that led to consistent increases in above-ground biomass in these fields.

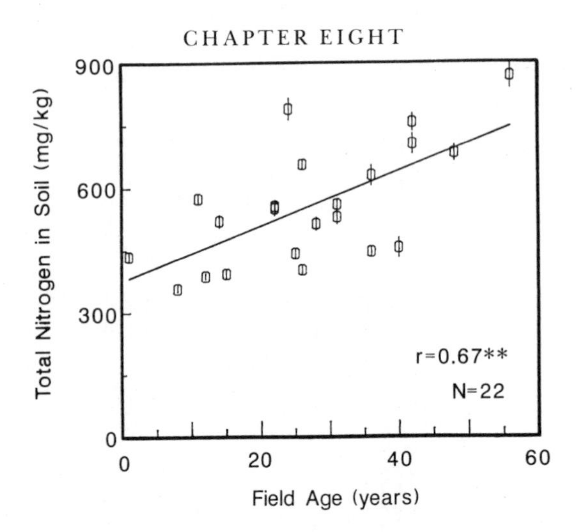

FIGURE 8.4. Total soil nitrogen (circles) increased significantly with field age, based on the chronosequence of 22 Cedar Creek old fields of known age since abandonment from agriculture. Standard errors are shown as vertical lines. Either 100 or 150 soil samples were analyzed per field. Figure from Tilman (1987a).

sion at Cedar Creek is an increase in the supply rate of nitrogen through time. As nitrogen supply increases, total plant biomass increases, light penetration to the soil surface decreases, and the level of available nitrogen in the soil increases, at least in the oldest fields.

The increases in extractable (available) nitrogen and the decreases in light penetration in the older, more nitrogen-rich fields is just what is predicted to occur if plant species are differentiated in their abilities to compete for nitrogen and light and if this differentiation is a cause of succession (Tilman 1982, 1985; Fig. 7.1). It is difficult to explain the higher levels of available nitrogen in the more nitrogen-rich fields if some other resource, such as light, were not also limiting. If nitrogen were the only limiting resource, species that could survive at the lower levels of available nitrogen in younger fields should increase in abundance in the older fields (with their greater levels of available nitrogen) and continue increasing in abundance until they reduced their

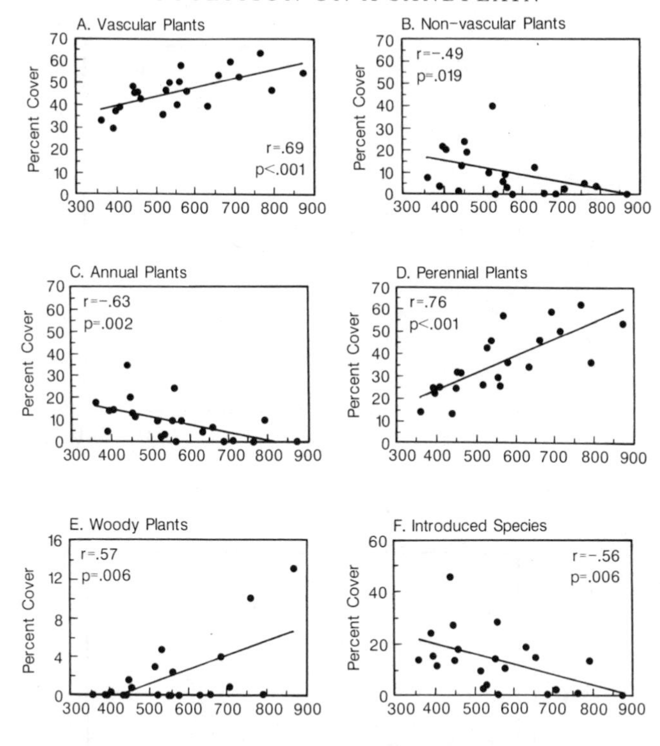

FIGURE 8.5. Abundances of various plant functional groups were significantly correlated with field average total soil nitrogen in the survey of 22 old fields at Cedar Creek. These data were first reported in Inouye et al. (1987a).

limiting resource, nitrogen, down to the lowest level at which they could survive (see Figure 2.1). This increase in available soil nitrogen is accompanied by a decrease in light penetration to the soil surface. It seems quite likely that light could be limiting as the penetration of light to the soil surface fell to less than 10% of incident sunlight. However, we have not performed any shading or light addition experiments in the field at Cedar Creek. Thus, the following discussion of the possible role of light at Cedar Creek must be considered tentative.

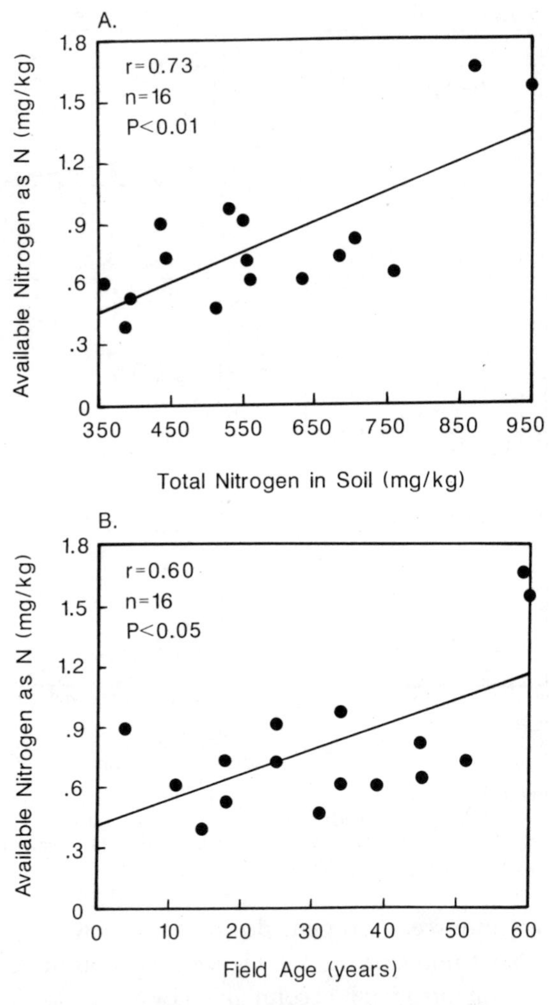

FIGURE 8.6. (A) Available nitrogen (the sum of 2 M KCl extractable ammonium and 2M KCl extractable nitrate) in 16 Cedar Creek old fields increased significantly with field average total soil nitrogen, with most of the increase occurring on the richest soils and oldest fields. It is the two oldest fields that cause these regressions to be significant. (B) It also increased with field age. Each point shows field average total nitrogen or field average available nitrogen for a particular field. Unpublished data collected by D. Tilman in the summer of 1986.

TESTABLE PREDICTIONS OF THEORY

Our survey of the soils and vegetation of Cedar Creek old fields suggests that plant competition for nitrogen and light, as well as time (where time may represent transient dynamics and/or colonization), may explain much of the successional pattern. Assuming that nitrogen and light are the two most important limiting resources at Cedar Creek, the theory developed in this book makes the following predictions:

(1) An increase in the supply rate of nitrogen should lead to an increase in total plant biomass and to a decrease in light penetration to the soil surface.

(2) A decrease in the supply rate of light, such as by shading, should lead to a decrease in total plant biomass and to an increase in the level of available soil nitrogen.

(3) Any change in the availability of nitrogen or light should lead to a change in the relative abundances of the plant species.

(4) The initial response to nitrogen addition should be a period of transient dominance by species with lower root:shoot ratios, i.e., with higher maximal growth rates than the eventual dominants.

(5) Eventually, plants should become separated along an experimental nitrogen gradient, with the order of occurrence along this gradient being the same as their order of occurrence along natural nitrogen gradients.

(6) The species that are dominant at the more nitrogen-rich end of either natural or experimental gradients should be taller at maturity and have a greater allocation to stems and a lower allocation to roots than species dominant at the lower nitrogen end.

(7) Initially different plant communities should converge on similar compositions if their rates of nitrogen supply are made similar.

(8) When growing in monoculture on nitrogen-poor soils, the species dominant at the nitrogen-poor end of a

productivity gradient should grow more rapidly at low nitrogen levels and should reduce available soil nitrogen to lower levels, at equilibrium, than the species dominant at the high-nitrogen end of the gradient.

(9) Once equilibrium is reached, the addition of nitrogen should lead to the same qualitative changes in the relative abundances of plant species as does the removal of light. Similarly, the addition of light should lead to the same qualitative changes as the removal of nitrogen.

EXPERIMENTAL TESTS

We have performed a series of experiments designed to test many, but by no means all, of these predictions. One of the main experiments consisted of seven separate experimental nitrogen gradients (Figs. 8.7 and 8.8; Tilman 1987a). Each gradient had 5 or 6 replicates of 9 different treatments. The treatments consisted of a control (treatment *I*) and 8 treatments that all received the same background mixture of P, K, Ca, Mg, S, and trace metals, but differed in the annual rate of nitrogen addition (Fig. 8.9). These 8 treatments, labeled *A* through *H*, received from 0 to 27 $g\ m^{-2}\ yr^{-1}$ of nitrogen, with treatment *A* receiving no nitrogen and treatment *H* receiving the highest annual rate of nitrogen addition. The experiments were established in 1982 in three successional fields that were 14, 25, and 48 years post-agriculture in 1982, and in native oak savanna. In each of the successional fields, two different experimental nitrogen gradients were established, one on the existing old field vegetation and one on an area that was newly disturbed via thorough disking in the spring of 1982. A single experimental nitrogen gradient was established on the existing vegetation in a prairie opening in the stand of native oak savanna. Each of these gradients was contained within a mammalian herbivore exclosure (Tilman 1987a).

FIGURE 8.7. An aerial photograph taken in 1983 of the various experimental plots in the 14-yr-old field, Field *A*. The smaller plots are 4 m × 4 m and are separated by 1 m walkways. The larger plots, partly visible in the middle left, are 20 m × 50 m. The main difference among these plots is the rate of nitrogen addition. The two experimental nitrogen gradients discussed in this book are inside the fence. The uppermost set of 54 plots were disturbed in 1982 before being divided into plots and fertilized. The lowermost set of 54 were not disturbed, but were also fertilized starting in 1982. Both sets are fenced to exclude mammalian herbivores. Note that this field is surrounded by forest, but that very few trees have become re-established in this unburned field. The field has, on average, about 40% of the total soil nitrogen of the surrounding, undisturbed forest.

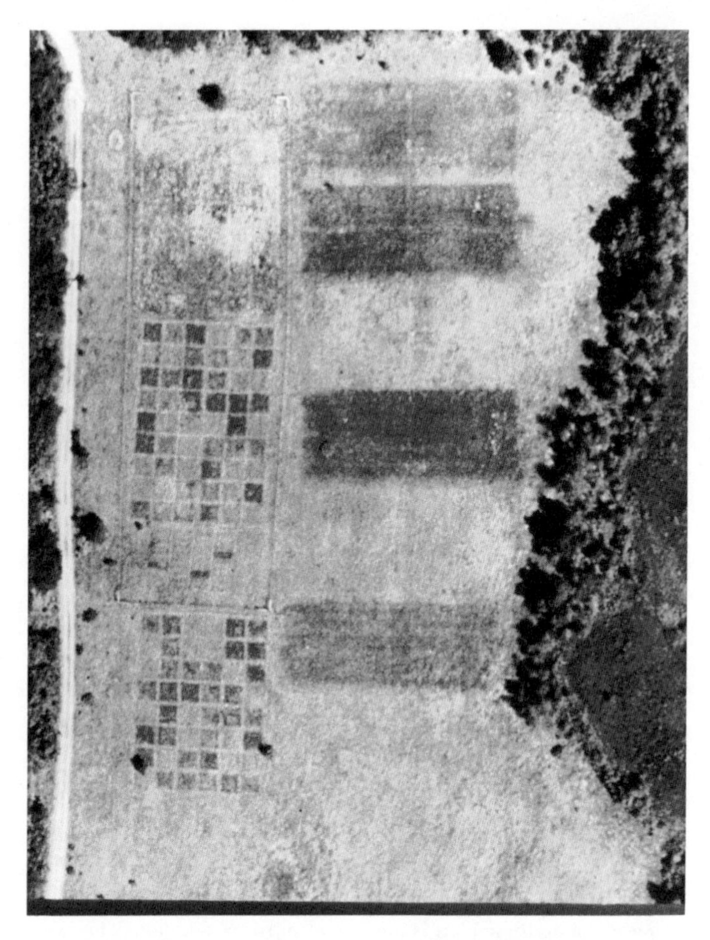

FIGURE 8.8. Aerial photograph taken in 1983 of the plots in the 48-year-old field, Field *C*. Smaller plots are 4 m × 4 m, and larger plots are 20 m × 50 m. Note that few trees have become re-established in this field, which has, on average, about 65% of the total soil nitrogen of the surrounding forest.

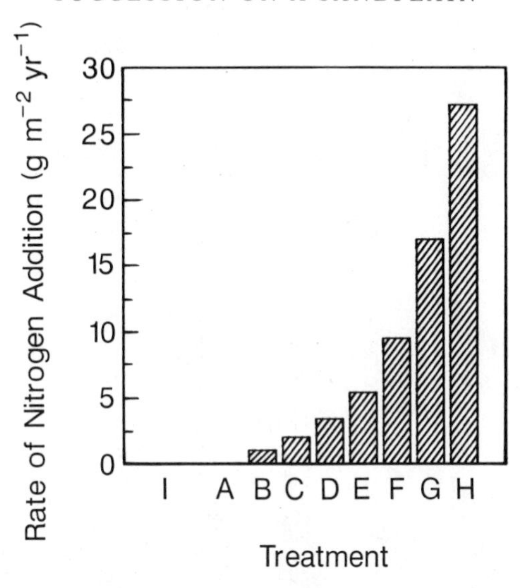

FIGURE 8.9. Nitrogen addition rates for the 9 treatments of the nitrogen gradient field experiments. The order in which these treatments are shown here is the same order in which responses to these treatments will be shown in the following figures. Note that treatment *I* received no nutrients of any kind, but that treatment *A* through *H* all received the same mixture of P, K, Ca, Mg, S, and trace metals. Treatment *A* received no nitrogen and treatment *H* received the highest rate of nitrogen addition. Treatments *A* through *H* thus form an experimental nitrogen gradient, with *I* being the control. See Tilman (1987a) for further details.

Biomass, Light and Nitrogen

On both the disturbed and the undisturbed gradients, in all fields, the responses to the experimental nitrogen gradients had many similar elements. In each year from 1982 through 1986, above-ground plant biomass increased (Fig. 8.10), light penetration to the soil surface decreased (Fig. 8.11), and vegetation height increased from the low to the high ends of the seven gradients (Fig. 8.12). Thus, these experiments showed that nitrogen was a limiting resource, and that increased supply rates of nitrogen led to increased plant biomass, and thus to decreased availability of light.

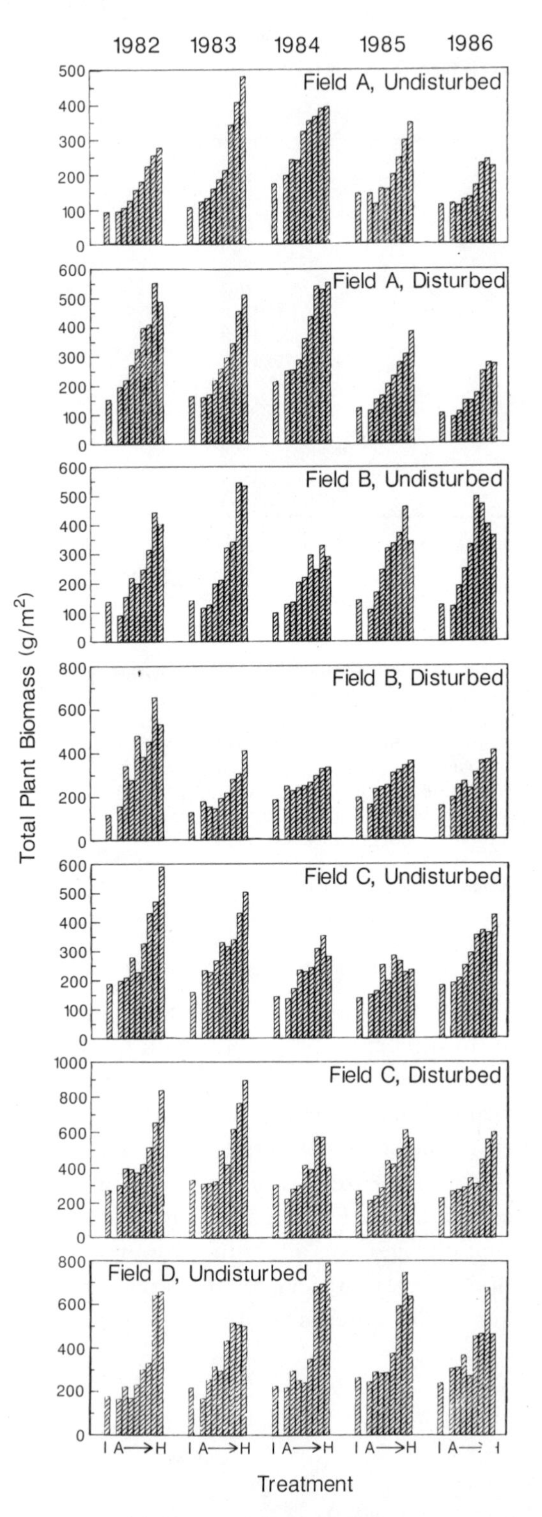

FIGURE 8.10. The dependence of total above-ground plant biomass on the experimental rate of nitrogen addition for each of the 7 experimental gradients in each year from 1982 through 1986. Each bar in each histogram is based on the average response of the six replicates of that treatment. The first column of histograms shows biomass responses along each nitrogen gradient in 1982, the next column shows those for 1983, etc. Field *A* was 14 yr old, *B* was 25 yr old, *C* was 48 yr old, and *D* was native savanna at the start of this work in 1982. The undisturbed gradients were established on existing vegetation. The disturbed gradients were established on soil that was thoroughly disked at the start of the experiment in 1982. Treatments are arranged in same order as shown in Figure 8.9, i.e., from low (*A*) to high (*H*) rates of nitrogen addition.

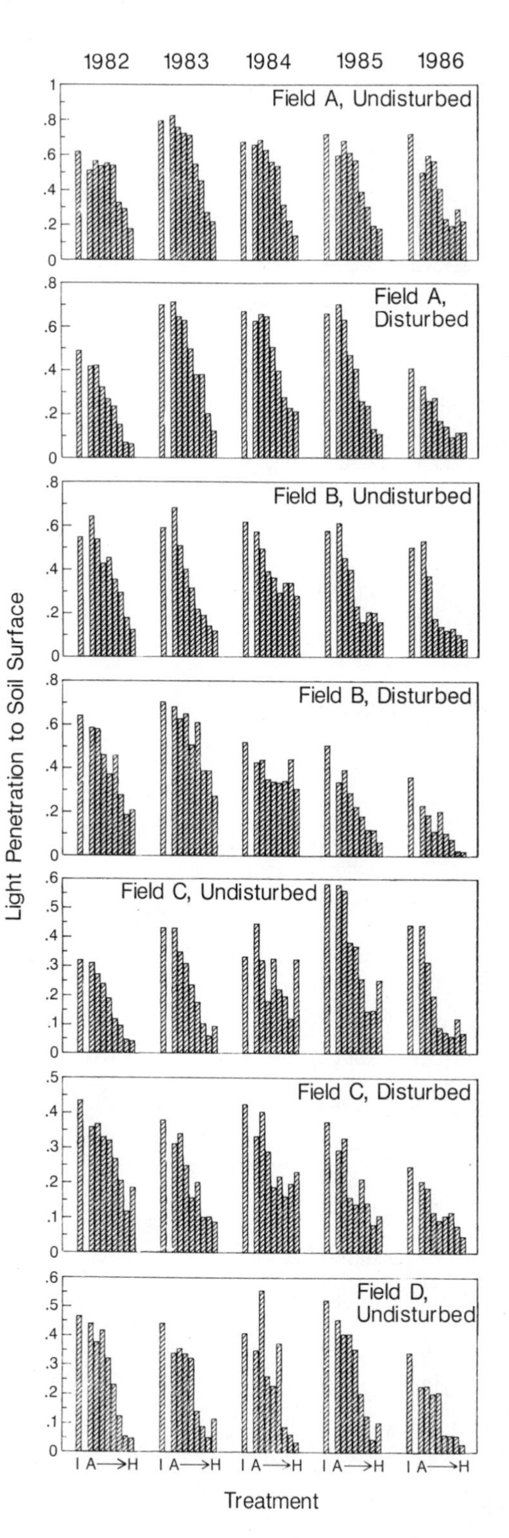

FIGURE 8.11. The dependence of light penetration to the soil surface on the experimental rate of nitrogen addition for each of the 7 Cedar Creek experimental gradients in each year from 1982 through 1986. The first column of histograms shows the dependence of light penetration on the nitrogen treatments in 1982, the next column shows those for 1983, etc. The undisturbed gradients were established on existing vegetation. The disturbed gradients were established on newly disked soil. Light penetration to the soil surface is expressed as the proportion of incident photosynthetically active radiation that reaches the soil surface.

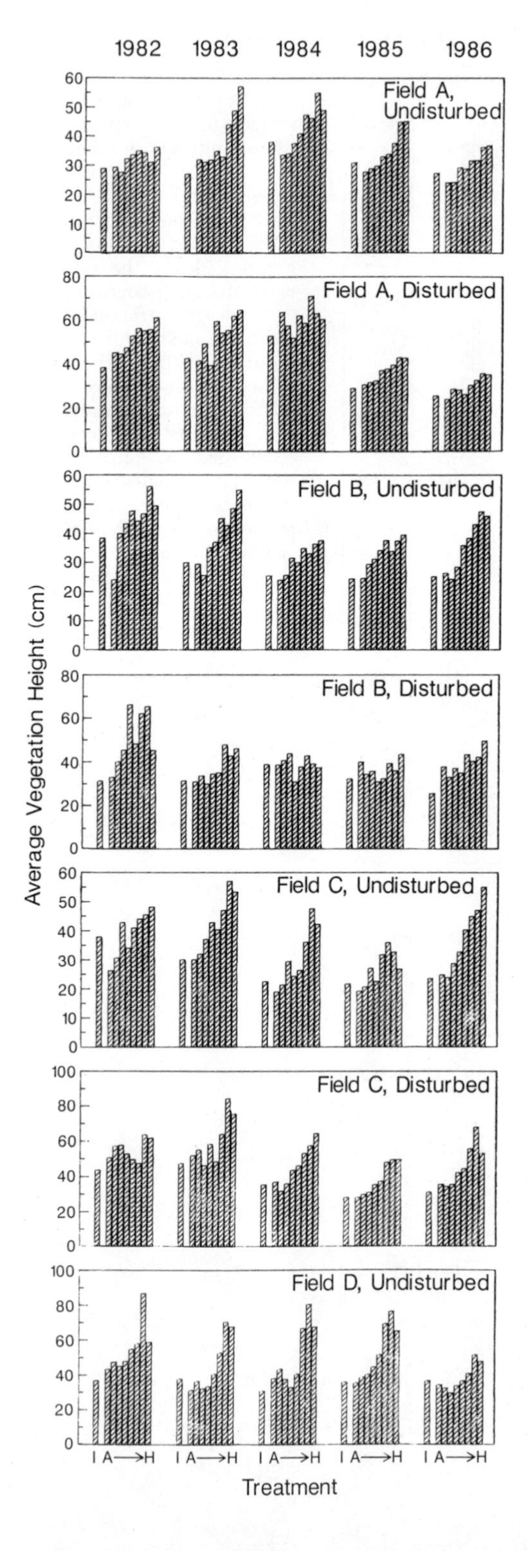

FIGURE 8.12. The dependence of average vegetation height on the experimental rate of nitrogen addition for each of the 7 Cedar Creek experimental gradients in each year from 1982 through 1986.

These changes, and the concomitant change in vegetation height, are consistent with predictions 1 and 6, above.

Another critically important prediction of this theory is that the level of available nitrogen should increase along these nitrogen supply gradients. Clearly, available nitrogen levels should be higher immediately after the periods of nitrogen addition in early and late spring each year. However, theory predicts that levels should remain higher throughout the growing season. To test this, the soils in each plot of each experimental nitrogen gradient were sampled at the time of peak biomass in mid to late summer of 1986. Available NH_4 + NO_3 was sampled by 2 M KCl extraction in the field, immediately after a soil sample was collected. Available nitrogen increased significantly with the rate of nitrogen supply on each of these seven experimental gradients, based on linear regressions. On average, over all seven gradients, available nitrogen (as N) in the soil increased from 0.4 mg/kg in the plots receiving no added nitrogen to 5.1 mg/kg in the plots receiving the highest rate of nitrogen addition. As predicted, available nitrogen and light available at the soil surface were inversely correlated for all 7 gradients. Two typical examples, for the undisturbed plots in the 25-year-old field and the native oak savanna, are shown in Figure 8.13. On either a gradient-by-gradient basis, or when all 7 gradients were averaged together (Fig. 8.14), increased rates of nitrogen addition led to higher levels of available nitrogen and to lower penetration of light to the soil surface.

Species Responses to Gradients

The responses of the different plant species to these experimental gradients provide a further test of this theory. From the first year of the experiments on, for each of the seven different gradients, there have been highly significant effects of the rate of nitrogen addition on the absolute and relative abundances of the common plant species (those with >2% of total plant biomass). During the first year of the

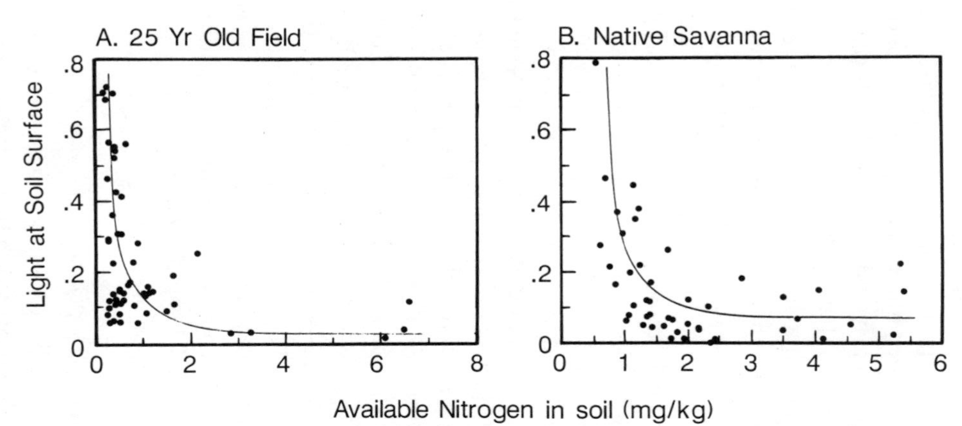

FIGURE 8.13. (A) Relation between light penetration to the soil surface and 2 M KCl extractable nitrogen (on a plot by plot basis) for the undisturbed experimental nitrogen gradient in the 25-yr-old field in the summer of 1986. (B) Similar data collected in the experimental nitrogen gradient on undisturbed native savanna in the summer of 1986.

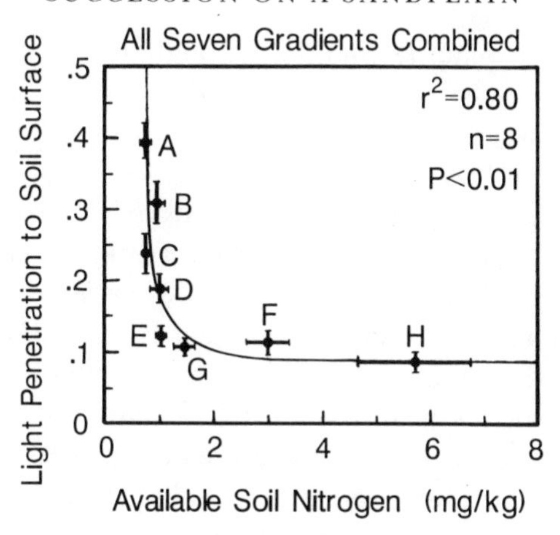

FIGURE 8.14. Relation between light penetration to the soil surface and 2 M KCl extractable nitrogen, based on the combined data from all 7 experimental nitrogen gradients. The letters *A* through *H* refer to treatments *A* through *H*. Treatment *I* was included with treatment *A*, because the two did not differ significantly. Each data point shows the mean and standard error (as bars) for both light and available (extractable) nitrogen for each treatment, based on 40 plots per treatment (except *A*, which has 80). Treatments *A* through *H* are the experimental nitrogen gradient (see Fig. 8.9), with *A* and *I* receiving no nitrogen and *H* receiving nitrogen at the greatest rate.

experiment, when plots were sampled approximately 2 to 3 months after the initial fertilizations, 29 of the total of 57 common species on the 7 gradients had their biomass increase significantly with the rate of nitrogen addition (Tilman 1987a). Not one of the common species had a significant decrease (based on linear regression) in its absolute abundance along these gradients during the first year. Thus, about half of these species were limited by nitrogen.

Although there is no direct experimental evidence showing what limited the other half of the species, there is much evidence showing what did not limit them. These gradients included plots that received no nutrients (treatment *I*) and plots that received all nutrients except nitrogen (treatment *A*). ANOVA's with contrasts between these two treatments, performed on a species-by-species basis, for each of the years from 1982 through 1985, showed that there were only 7 instances out of the total of 232 possible, for which a species had higher biomass after the addition of a mixture of P, K, Ca, Mg, S, and trace metals (Cu, Co, Mo, Zn, and Mn) compared to the unfertilized controls (Tilman 1987a). Each of these 7 cases was for a different species in a different field or year. These 7 cases are most likely Type I errors. These data demonstrate that P, K, Ca, Mg, S, and trace metals are unlikely to limit the growth of any of the common species in any of these seven gradients. This is consistent with results of experiments in which various resources including water were added singly to replicate plots in these same four experimental fields (Fig. 8.3). Thus, there is strong experimental evidence that the species that were not limited by nitrogen were also not limited by any other soil resource, including water.

Some other lines of evidence suggest that light may have been the other major limiting resource along these gradients. Increased rates of nitrogen supply decreased light penetration to the soil surface to less than 10% of incident light, on average, compared to 40% reaching the soil surface in the unfertilized control plots (Fig. 8.14). At this level of irradiance, seedlings and shoots of newly establishing plants could be light limited (Bazzaz 1979). Further, there was a strong tendency for the species that attained greater heights early in the growing season to be the dominants of the high nitrogen plots.

Through time, the common species, especially the perennials, became increasingly distinct in their distributions along the 7 experimental nitrogen gradients (Tilman 1987a). Further, each perennial species became increasingly similar in its responses along the 7 gradients, independent of its initial abundances on each gradient (e.g., Fig. 8.21). The annuals also became differentiated along the gradients, but their responses were more variable from year to year. The youngest field, which had been abandoned from farming for 14 years when this work began, had many annuals and a few perennials in 1982. From 1984, and on, the native annuals, *Ambrosia artemisiifolia* (ragweed) and *Hedeoma hispida* (a mint), attained their peak relative abundance at the lowest nitrogen end of the disturbed (disked) and undisturbed nitrogen gradients in this field, and declined along the gradient (Tilman 1987a). Some of the introduced annuals, such as *Berteroa incana* (hoary alyssum), *Polygonum convolvulus* (bindweed), and *Erigeron canadensis* (horseweed), attained their peak relative abundance at intermediate points on the gradients. By 1986, two perennial grasses completely dominated both the disturbed and the undisturbed gradients in this young field: *Agropyron repens* (quack grass) at the high nitrogen ends of the gradients and *Poa pratensis* (Kentucky bluegrass) at the low nitrogen ends.

The few annuals that were common in the 25-year-old field occurred on the disturbed gradient, and were rare by 1986. The most abundant perennials, *Schizachyrium scoparium* (little bluestem), *Poa pratensis*, and *Agropyron repens*, responded similarly to both the undisturbed and the disturbed nitrogen gradients. *Schizachyrium scoparium* declined along both gradients in the 25-year-old field. *Poa pratensis* reached its peak abundance at intermediate rates of nitrogen supply. High nitrogen plots were dominated by *Agropyron repens*.

A similar pattern occurred in the 48-year-old field by the fifth year of this experiment. *Schizachyrium scoparium* dominated low nitrogen treatments, *Poa pratensis* and *Panicum oligosanthes* dominated intermediate regions of the gradients, and *Agropyron repens* dominated the high nitrogen treatments on both the undisturbed (Fig. 8.15) and the disturbed gradients (Tilman 1987a). In addition, several other common species had distinct distributions along the gradients. *Solidago nemoralis* (a rosette-forming goldenrod) was most abundant in low nitrogen treatments, while *Artemisia ludoviciana* (a prairie sage) was most abundant in intermediate treatments (Fig. 8.15).

The prairie opening in the native oak savanna had a single gradient, on undisked vegetation. The common species became distinctly separated, with *Sorghastrum nutans* (Indian grass) reaching its peak abundance at the low nitrogen end (Fig. 8.15). The intermediate rates of nitrogen addition were dominated, in order from lower to higher rates, by the legume *Lathyrus venosus*, *Panicum perlongum*, *Poa pratensis*, *Ambrosia coronopifolia*, and then *Artemisia ludoviciana*. The highest rates of nitrogen supply were dominated by the woody vine, *Rubus* sp. (blackberry) and by the goldenrod, *Solidago graminifolia* (Fig. 8.16).

In total, these data clearly demonstrate that species become differentiated along experimental nitrogen gradients. Further, they show that species tend to respond similarly to the gradients, despite great initial differences in their abundances. For instance, *Agropyron repens* was rare or absent from many plots in the 25- and 48-year-old fields, and from all plots in the oak savanna. It invaded and then increased rapidly in abundance in the high nitrogen plots in the 25- and 48-year-old fields, having qualitatively similar responses to nitrogen addition in 6 of the 7 gradient experiments. The only gradient along which it did not increase dramatically with the rate of nitrogen addition was that in the native savanna, on which it did not occur. *Poa pratensis*, which occurred on all 7 gradients, reached its peak abundance at low to intermediate rates of nitrogen addition. In the absence of *Schizachyrium scoparium* in the youngest field, it attained its peak abundance at the lowest rates of nitrogen supply. In the presence of *Schizachyrium*, its peak was moved to intermediate rates of nitrogen supply, and *Schizachyrium* dominated the plots with low rates of supply. Many other comparisons of species responses along these gradients are given in Tilman (1987a). With a very few exceptions, each species had converged, by 1986, on the same abundance pattern along all experimental gradients on which it was present, independent of its initial rarity or abundance in 1982.

Grime (1979, p. 16) proposed what he called a "unified concept of competitive ability" in which "the abilities to compete for light, water, mineral nutrients and space" all vary in concert. He suggested that some species are superior competitors for all resources (e.g., nitrogen, water, and light) and other species are poor competitors for all resources. If Grime were correct, plants would not become separated along experimental resource gradients, such as the nitrogen gradients at Cedar Creek.

Thus, these experiments provide strong evidence that refutes Grime's (1979) view. The separation of species along these experimental gradients is consistent with the hypothesis that they are differentiated in their competitive abilities for nitrogen and light. However, if that hypothesis is correct, then species with shorter stature and shorter life spans should be dominant at lower rates of nitrogen supply, and taller, longer-lived perennials should be dominant at higher rates of nitrogen supply. Although the pattern is by no means perfect, there has been a consistent tendency for these experiments to support this prediction (Tilman 1987a). Plant height at maturity does tend to increase with the rate of nitrogen addition (Figs. 8.15, 8.16). Rank order correlations between height at maturity and position of peak abundance of each of the common species on each gradient in 1984 showed that 6 of the 7 correlations were positive, as expected, with two of these significant at $P < 0.05$. Similar analyses using the 1985 data showed that all 7 correlations were positive, as expected, again with two of them significant (Tilman 1987a). Although this is a low proportion of significant correlations, on average there were only 6 common species per gradient, and thus only very strong trends were significant.

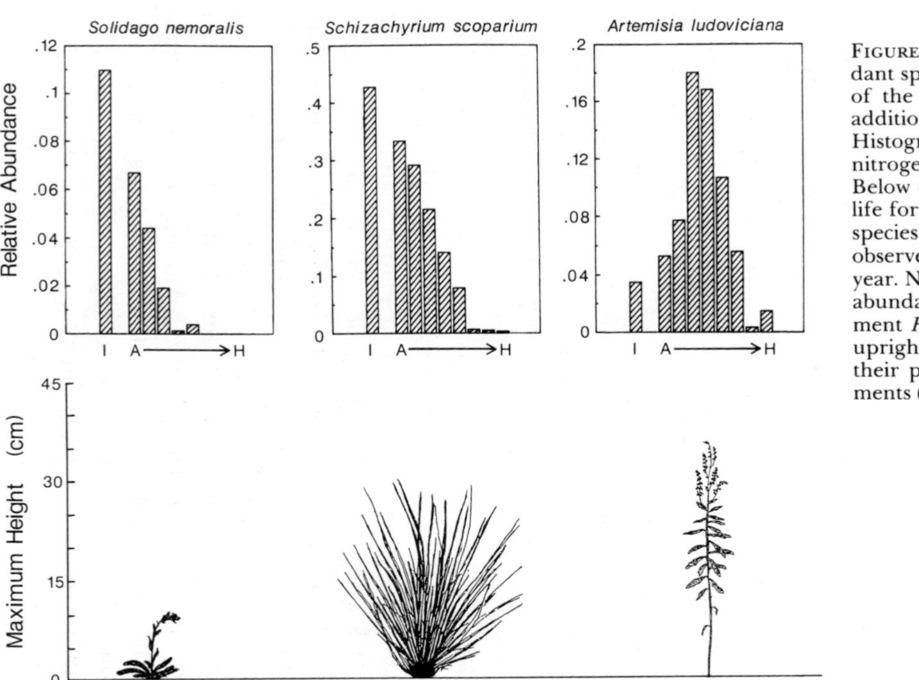

FIGURE 8.15. The responses of the 6 most abundant species of the undisturbed nitrogen gradient of the 25-yr-old field to experimental nitrogen addition in 1986, the fifth year of the experiment. Histograms show relative abundance versus nitrogen addition treatment, as in Figure 8.9. Below each histogram is a sketch, to scale, of the life form of the plant. The height shown for each species is its maximal treatment-average height observed along the experimental gradient that year. Note that the species that reached their peak abundance in the high nitrogen treatments (treatment *H*) are taller at maturity and have a more upright growth form than the species that reached their peak abundance in the low nitrogen treatments (treatment *A*).

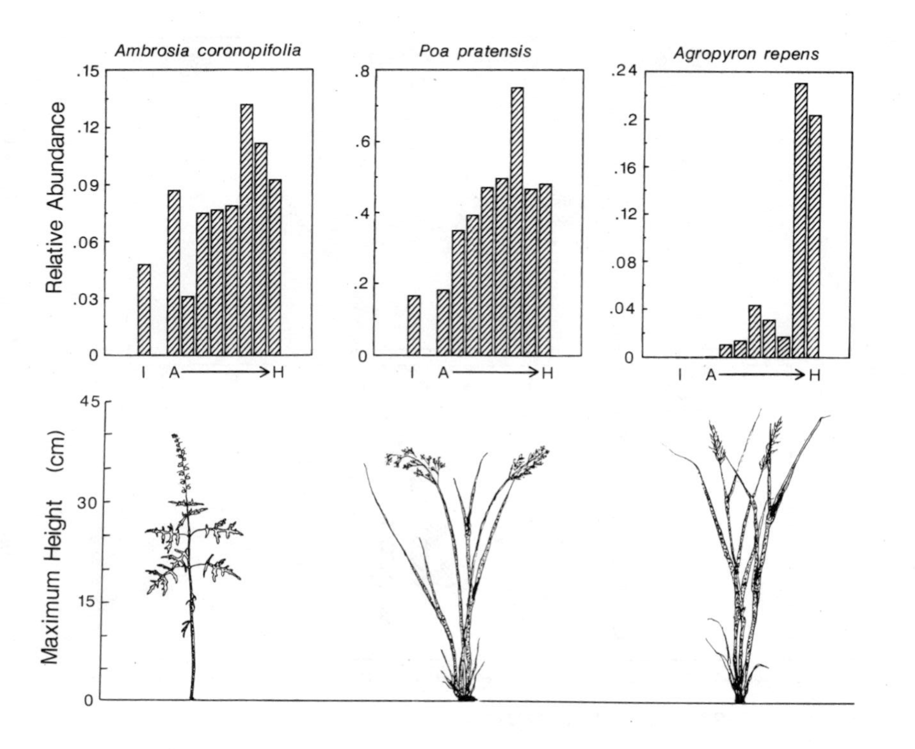

Ambrosia coronopifolia
Poa pratensis
Agropyron repens
Relative Abundance
.15
.12
.09
.06
.03
0
.8
.6
.4
.2
0
.24
.2
.16
.12
.08
.04
0
I A → H
Maximum Height (cm)
45
30
15
0

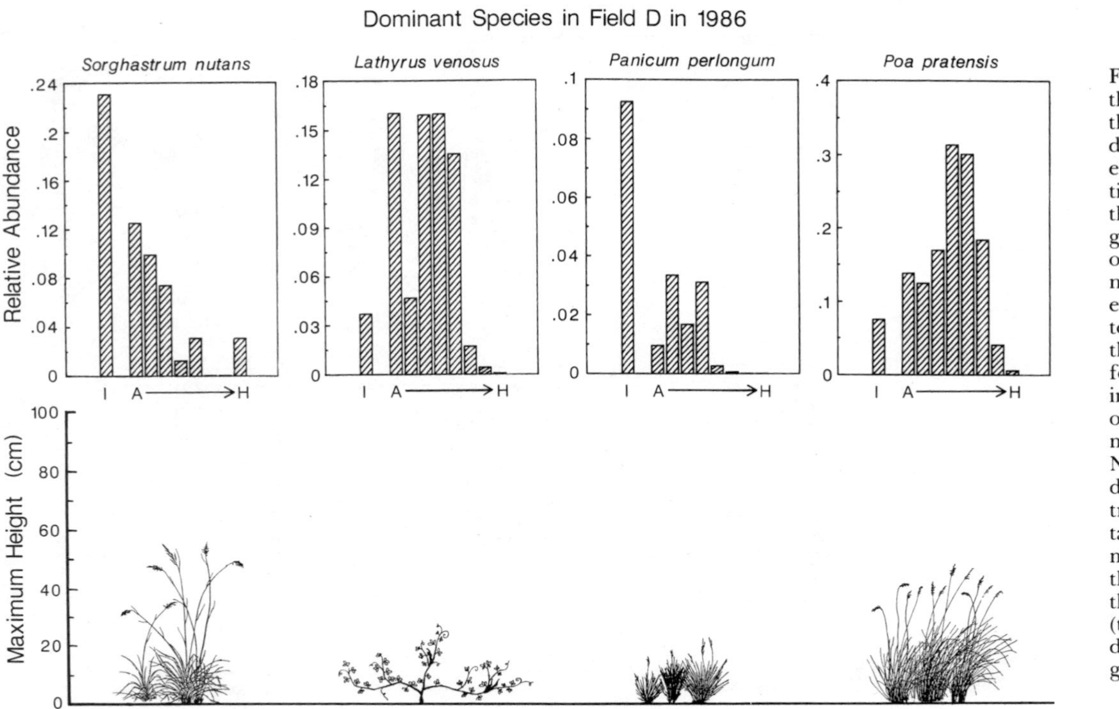

FIGURE 8.16. The responses of the 8 most abundant species of the undisturbed nitrogen gradient of the native savanna to experimental nitrogen addition in 1986, the fifth year of the experiment. Each histogram shows the dependence of relative abundance on the nitrogen treatments. Below each histogram is a sketch, to scale, of the life form of the plant. The height shown for each species is its maximal treatment-average height observed along the experimental gradient that year. Note that the species that dominated the high nitrogen treatments (treatment *H*) are taller at maturity and have a more upright growth form than the species that dominate the low nitrogen treatments (treatment *A*). *Rubus*, which dominated the highest nitrogen plots, is a woody vine.

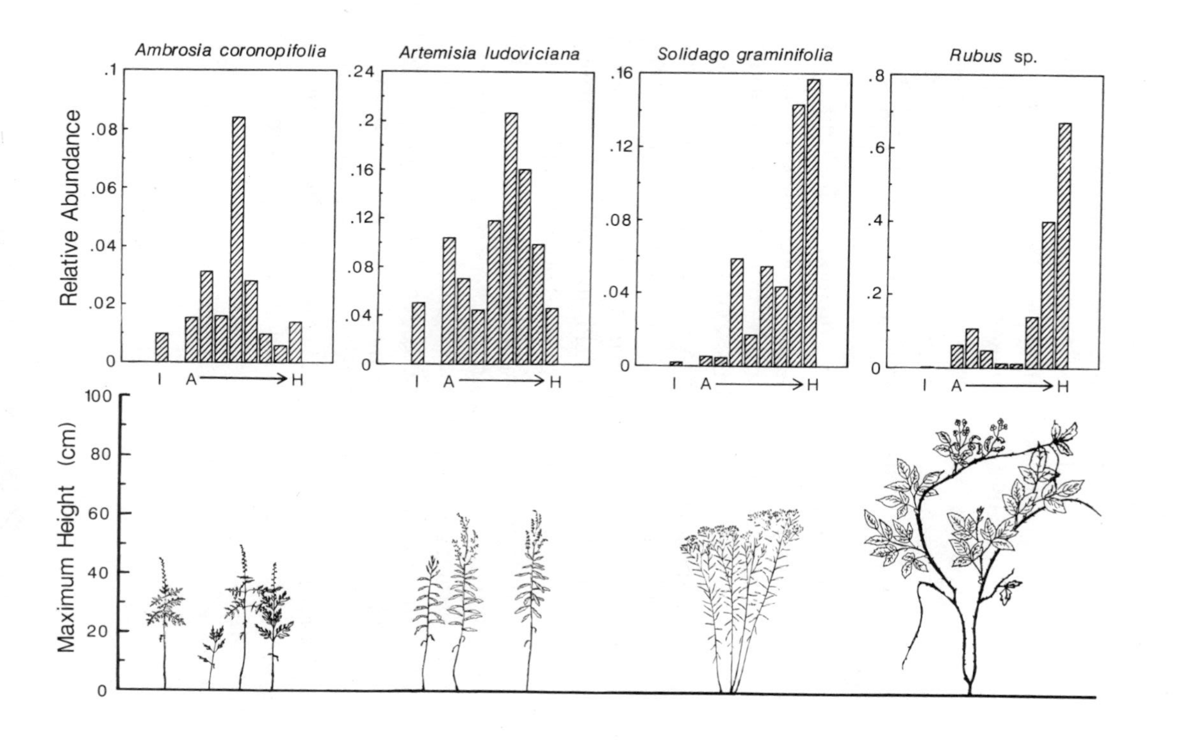

Ambrosia coronopifolia
Artemisia ludoviciana
Solidago graminifolia
Rubus sp.
Relative Abundance
.1
.08
.06
.04
.02
0
.24
.2
.16
.12
.08
.04
0
.16
.12
.08
.04
0
.8
.6
.4
.2
0
I A → H
Maximum Height (cm)
100
80
60
40
20
0

These experimental results demonstrate that nitrogen is the major limiting soil resource and that different species reach their peak abundances at different points along experimental nitrogen gradients. How, though, do the patterns of species abundances along these experimental gradients compare to their responses to point-to-point nitrogen spatial heterogeneity within individual fields or to differences in soil nitrogen among successional fields? If nitrogen is an important determinant of the natural distributional patterns of these species and if the responses to the experimental gradients by 1986, the fifth year of the experiment, are indicative of the long-term effect of the experimental gradients, then species responses to the experimental nitrogen gradients should be consistent with correlational patterns observed within and among these fields.

Of all the species that were common in the 7 different experimental nitrogen gradients, 15 were also among the 20 most abundant species in the old field survey (Tilman 1987a). By the fifth year of the experiment, most of these species responded similarly to the spatial nitrogen variability within individual fields and to the 7 experimental nitrogen gradients. For instance, *Agrostis scabra*, bent grass, occurred in 21 of the 22 old fields in which we performed our survey of soils and vegetation. For each field, I determined if the percent cover of *Agrostis* within each 0.5 × 1.0 m quadrat was correlated with the total soil nitrogen level of the quadrat (as estimated by a single sample collected at the center of each quadrat). There were 100 or 150 quadrats per field. The abundance of *Agrostis* was negatively correlated with soil nitrogen in 17 fields, with 8 significant at $P < 0.05$. Six of these are shown in Figure 8.17. There were positive correlations in 4 fields, but none were significant at $P < 0.05$. Indeed, the most positive had a correlation coefficient of just $r = 0.06$. As Figure 8.17 clearly illustrates, there is much variance in the abundance of this species that is not explained by soil nitrogen. The most significant correlation explained just 25% of the plot-to-plot variance in a field for *Agrostis*. However, despite such small-scale variability, *Agrostis scabra* had a consistent tendency to be more abundant in the nitrogen-poor areas of these fields. This same pattern held when all fields were considered together (Fig. 8.18). Using all the data from all 21 fields in which *Agrostis* occurred, its cover was highly significantly negatively correlated with soil nitrogen ($r = -0.28$, $n = 2162$, $P < 0.001$). A summary curve, created by dividing all the data into 40 different nitrogen ranges, and calculating the cover of *Agrostis* within each, shows that it reached its peak abundance in the old field survey in very nitrogen-poor areas, and declined with increasing nitrogen levels (Fig. 8.18).

This pattern of *Agrostis scabra* reaching its peak cover in quadrats and fields with low nitrogen and being less abundant in areas with more nitrogen is consistent with its response to nitrogen fertilization. From the second year of the experiment, and on, *Agrostis* has declined in abundance along the experimental nitrogen gradients in all fields in which it was common (Fig. 8.19). Thus, the response of *Agrostis* to experimental nitrogen addition is consistent with its natural distributional pattern both within spatially heterogeneous fields and among all fields in which it occurred.

Several other species tended to decline in abundance with increases in soil nitrogen within individual fields, and also declined in abundance along the experimental nitrogen gradients. For instance, *Hedeoma hispida*, a mint, occurred in 16 of the 22 fields, declined in cover with increases in soil nitrogen in 14 of these fields, and had significantly negative correlations in 9 of these fields. It increased in abundance with nitrogen in 2 fields, with one of these being a significantly positive correlation. It declined in abundance along the experimental nitrogen gradients in 1983, 1984, 1985, and 1986 for both gradients on which it was common

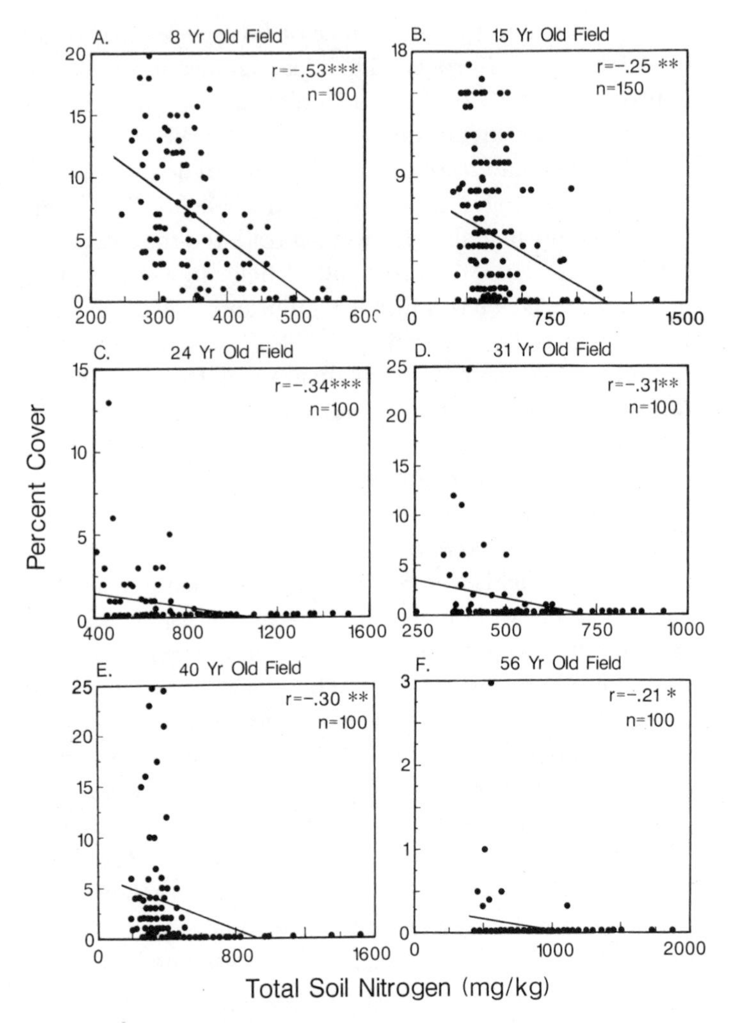

FIGURE 8.17. Each of these graphs shows the observed dependence of the percent cover of *Agrostis* on total soil nitrogen within a particular old field at Cedar Creek. Lines are regressions. Correlation coefficients, *r*, are shown. Each graph is based on estimates of vegetative cover and total soil nitrogen in 100 or 150 quadrats per field.

FIGURE 8.18. Histograms showing the dependence of percent cover of 15 common old field plants on total soil nitrogen in 22 Cedar Creek old fields. The histogram for each species used all the data for all fields in which a species was present. Each figure also shows whether there was a significant linear (*L*) or quadratic (*Q*) regression of percent cover on quadrat total soil nitrogen. Histograms were smoothed once using running averages. From Tilman (1987a).

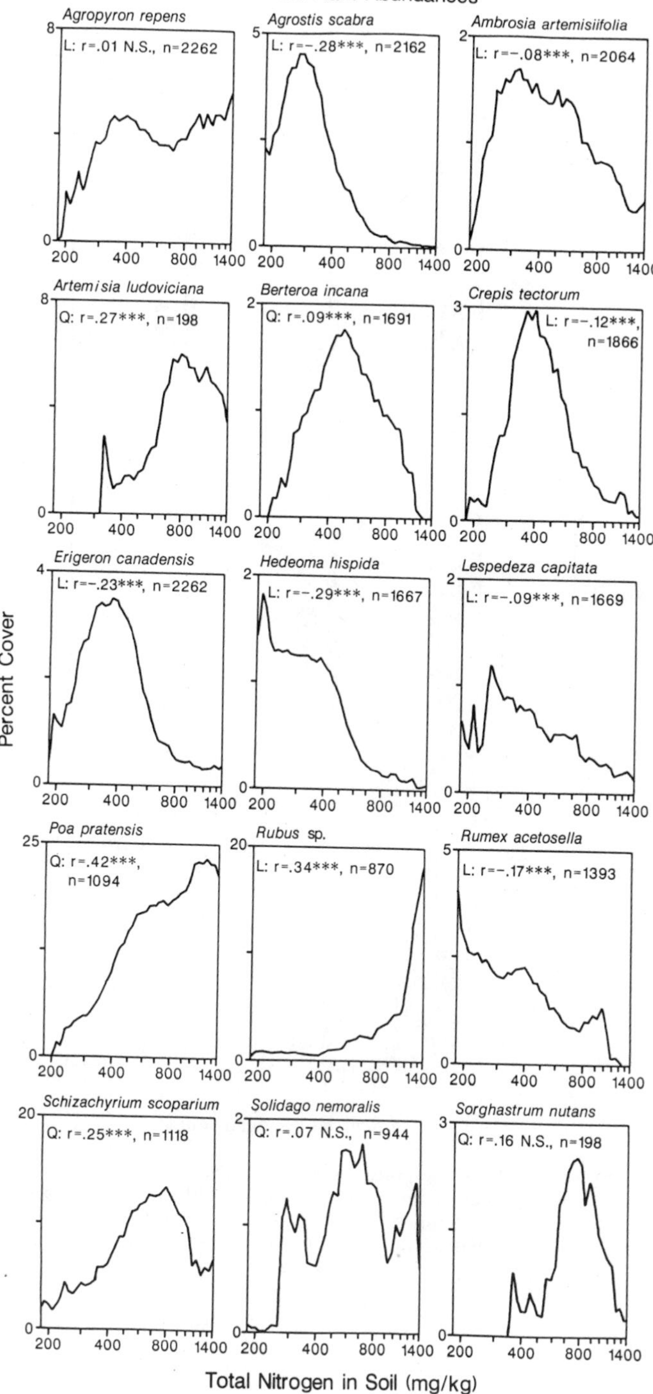

Old Field Plant Abundances
Agropyron repens
L: r=.01 N.S., n=2262
Agrostis scabra
L: r=-.28***, n=2162
Ambrosia artemisiifolia
L: r=-.08***, n=2064
Artemisia ludoviciana
Q: r=.27***, n=198
Berteroa incana
Q: r=.09***, n=1691
Crepis tectorum
L: r=-.12***, n=1866
Erigeron canadensis
L: r=-.23***, n=2262
Hedeoma hispida
L: r=-.29***, n=1667
Lespedeza capitata
L: r=-.09***, n=1669
Poa pratensis
Q: r=.42***, n=1094
Rubus sp.
L: r=.34***, n=870
Rumex acetosella
L: r=-.17***, n=1393
Schizachyrium scoparium
Q: r=.25***, n=1118
Solidago nemoralis
Q: r=.07 N.S., n=944
Sorghastrum nutans
Q: r=.16 N.S., n=198
Percent Cover
Total Nitrogen in Soil (mg/kg)

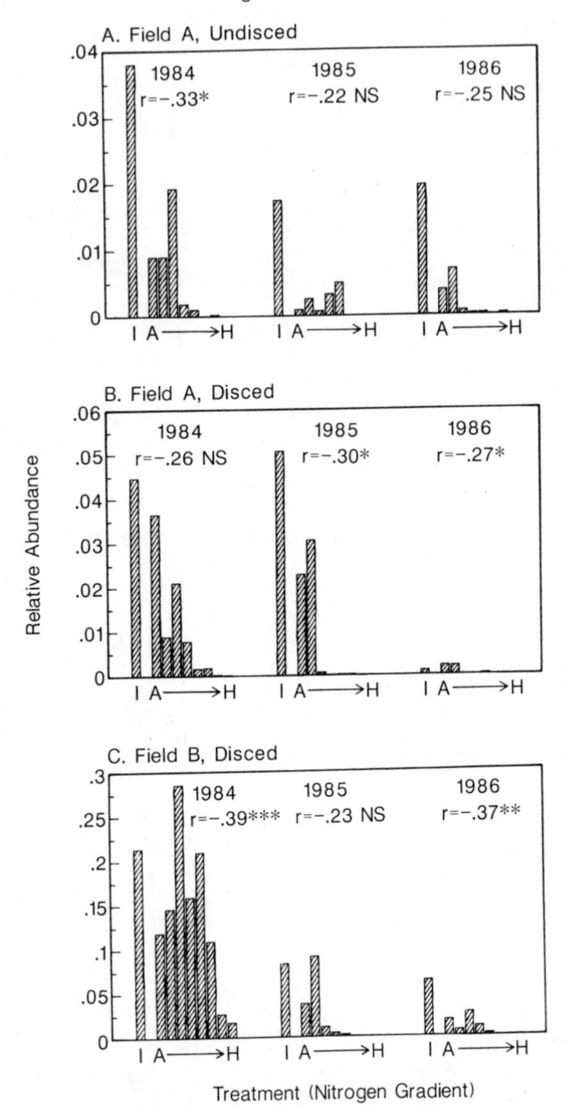

FIGURE 8.19. (A) Response of *Agrostis scabra* to the experimental nitrogen gradient in the undisked plots of Field *A* in 1984, 1985 and 1986. (B) Its responses in the disked plots of Field *A*. (C) Its response in the disked plots of Field *B*. It was rare or absent in the

(Tilman 1987a). Neither *Ambrosia artemisiifolia*, *Crepis tectorum*, *Erigeron canadensis*, *Lespedeza capitata*, nor *Rumex acetosella* had any consistent correlation, from field to field, between its abundance and soil nitrogen. However, when the data for all fields were considered together, simple correlations showed that each of these decreased in abundance with total soil nitrogen with all of them reaching their peak abundance at low total soil nitrogen levels (Fig. 8.18). Each of these tended to decrease in abundance with added nitrogen on most of the gradients on which it was common, but there were several exceptions, especially for the annuals *Erigeron canadensis* and *Crepis tectorum* (Tilman 1987a).

Other species increased in abundance with soil nitrogen. For instance, *Agropyron repens*, which occurred in all 22 fields, had its cover positively correlated with soil nitrogen for 16 fields, with 8 of these significant at $P < 0.05$. Its cover was negatively correlated with nitrogen in 6 fields, of which only one was significant (Fig. 8.20). As already discussed, it had a consistent and highly significant pattern of dominating the high nitrogen plots of the 6 gradients on which it occurred. Thus, its dominance of plots receiving high rates of nitrogen addition is consistent with its pattern of being more abundant within the more nitrogen-rich areas of individual fields. However, when the data for all 22 fields were analyzed together, there was no significant correlation between soil nitrogen and the abundance of *Agropyron repens* (Fig. 8.18). This lack of a significant correlation came because *Agropyron* was less abundant in older fields, on average, than in younger fields. Even though it was consistently more abundant in the more nitrogen-rich portions of individual fields, it was apparently being displaced from fields as they aged, even though older fields were more

---

other 4 experimental nitrogen gradients. The correlation coefficients (r) shown are based on a simple correlation between the relative abundance of *Agrostis* and the rate of nitrogen addition, using all plots for a given year and gradient.

nitrogen rich. This contradicts the role of increasing soil nitrogen during succession as a determinant of *Agropyron*'s successional dynamics. As will be discussed, its early dominance may represent transient dynamics. *Agropyron* may be a species specialized on disturbed, high-nitrogen habitats.

Another species that increased in abundance with nitrogen within individual fields was the woody vine, the blackberry, *Rubus* sp. *Rubus* occurred in 9 fields. In 5 of these its cover was significantly positively correlated with soil nitrogen, in one it was a significant quadratic function of nitrogen, and in one it was significantly negatively correlated with nitrogen. When all the data from these 9 fields were analyzed as a unit, there was a highly significant positive correlation between its abundance and soil nitrogen, with its abundance increasing almost exponentially with soil nitrogen (Fig. 8.18). It was common only in the field of native oak savanna. Its relative and absolute abundance increased highly significantly along the experimental nitrogen gradient in this field from 1983 and on (Fig. 8.16; Tilman 1987a).

Another species that had greater abundance in more nitrogen-rich areas within individual old fields was *Poa pratensis*. It occurred in 11 of the 22 fields, and had its abundance positively correlated with nitrogen in 9 fields, with six of these correlations being significant. In addition, its abundance was a significantly quadratic (Gaussian) function of nitrogen in two fields. When data from all 11 fields in which it occurred were combined, and analyzed as a group, its abundance was a significantly quadratic function of total soil nitrogen (Fig. 8.18). This suggests that *Poa pratensis*

FIGURE 8.20. Dependence of the percent cover of *Agropyron repens* on total soil nitrogen for each of 8 different Cedar Creek old fields. These are the 8 fields with the highest abundances of *Agropyron* and with significant linear or quadratic correlations between its abundance and soil nitrogen. For each field, each point shows the percent cover of *Agropyron* in a 0.5 × 1.0 m quadrat and the total soil nitrogen for a 15 cm soil core collected at the center of that plot.

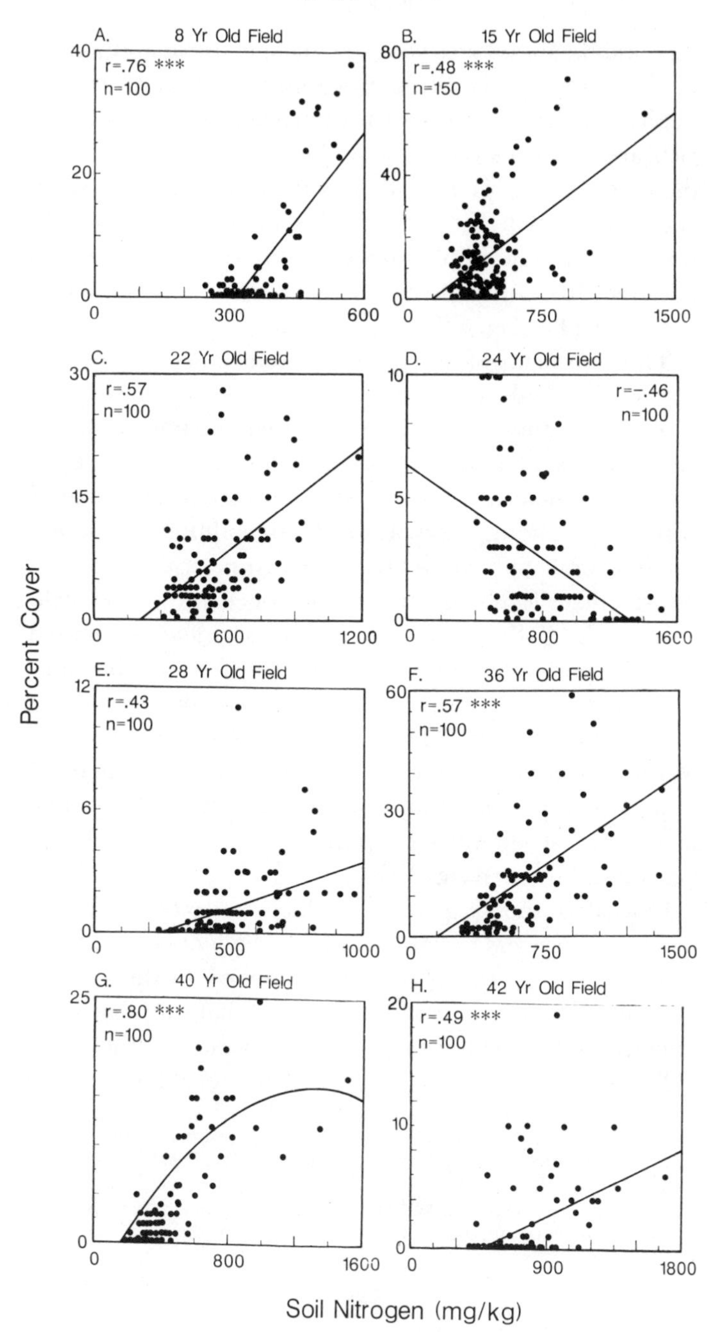

Agropyron repens
A. 8 Yr Old Field
r=.76 ***
n=100
B. 15 Yr Old Field
r=.48 ***
n=150
C. 22 Yr Old Field
r=.57
n=100
D. 24 Yr Old Field
r=-.46
n=100
E. 28 Yr Old Field
r=.43
n=100
F. 36 Yr Old Field
r=.57 ***
n=100
G. 40 Yr Old Field
r=.80 ***
n=100
H. 42 Yr Old Field
r=.49 ***
n=100
Percent Cover
Soil Nitrogen (mg/kg)

increases in abundance with soil nitrogen throughout most of the range of soil nitrogen within the Cedar Creek old fields, but that it declines with nitrogen in very nitrogen-rich areas. For its response to the nitrogen addition gradients to be consistent with this distributional pattern, it should increase in abundance at low rates of nitrogen addition, and decline in abundance at high rates of nitrogen addition. Just this pattern was observed in all of the fields except Field *A* (Fig. 8.21).

Thus, for many of the common species that tended either to increase with soil nitrogen or to decrease with soil nitrogen in the old field survey, their responses to the experimental nitrogen gradients support the role of nitrogen in determining their spatially patchy distributions within and among these old fields. In addition to *Poa pratensis*, several other species reached their peak abundance at intermediate levels of total soil nitrogen in the old field survey (Fig. 8.18). Of these, *Artemisia ludoviciana* responded to the experimental nitrogen gradients in a way that supported the role of soil nitrogen in determining its natural distribution pattern (Tilman 1987a). However, *Schizachyrium scoparium* did not. Its response to the experimental gradients (Fig. 8.15) would predict that it would decline along natural soil nitrogen gradients, but it actually had a Gaussian response (Fig. 8.18).

In total, of the 12 species for which percent cover in the old field survey was significantly dependent on total soil nitrogen (Fig. 8.18), 9 species responded to the experimental nitrogen gradients in a manner that supported the hypothesis that soil nitrogen was a major cause of their spatially patchy distributions within and among old fields.

FIGURE 8.21. Response of *Poa pratensis* to the experimental nitrogen gradients in the undisked plots of Field *B* (A), the disked plots of Field *B* (B), the undisked plots of Field *C* (C), the disked plots of Field *C* (D) and the undisked plots of Field *D* (E). Note the convergence on similar quadratic responses to the experimental nitrogen gradients.

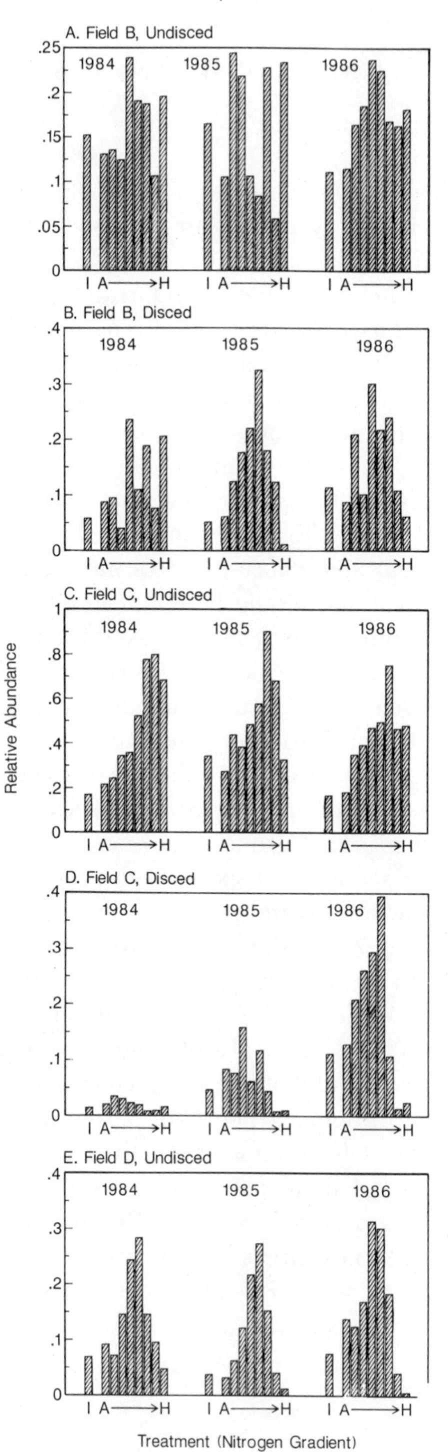
Poa pratensis
A. Field B, Undisced
1984
1985
1986
B. Field B, Disced
C. Field C, Undisced
D. Field C, Disced
E. Field D, Undisced
Relative Abundance
I A→H
Treatment (Nitrogen Gradient)

CHAPTER EIGHT

SUCCESSIONAL DYNAMICS

The theory presented in Chapter 7 predicted that, during secondary succession on a poor soil, the initial species sequence would be determined mainly by transient dynamics, and that slow increases in a soil nutrient would mainly influence the later stages of succession. The Cedar Creek experiments provide an opportunity to test this prediction in comparison with some alternatives. One alternative is that transient dynamics are unimportant and that the full successional sequence can be explained purely as resulting from increasing soil nitrogen and its effects on competition. If nitrogen accumulation were the main cause of the successional sequence at Cedar Creek, then early successional fields should be dominated by species that reach their peak abundance at low soil nitrogen levels, fields of intermediate age should be dominated by species that reach peak abundance at intermediate levels of nitrogen addition, and late successional fields should be dominated by species that reach peak abundance at high rates of nitrogen addition. Another alternative is that the full successional sequence is determined by transient dynamics. If this were so, the early successional species should have the greatest maximal growth rates, and these should be replaced by a sequence of species with progressively lower maximal growth rates and progressively greater allocation to stem (see Fig. 7.5). A third alternative is that the sequence is determined by colonization rates and competitive interactions. This would require that initial dominants either be present in the fields upon abandonment or have very well-dispersed seeds, and that late successional species be superior competitors but have poorly-dispersed seed.

In order to evaluate these alternatives, it is necessary to know the actual dynamics of succession at Cedar Creek. To determine this, I combined two different data sets. One data set was from the previously described survey of 22 old fields (Inouye et al. 1987a). The other data set was from a survey of eleven 20 m × 100 m strips of vegetation that have been sequentially abandoned from agriculture (one strip per year) from 1974 to 1984. This project was initiated by Don Lawrence and the data were collected by Barbara Delaney. I thank both of them, especially Barbara, who kindly allowed me to use unpublished data. The average abundance of each species was calculated for each field or each successional strip, and these 33 points were graphed against successional age. A smooth curve was then drawn for each species to summarize how its abundance depended on successional age. These curves are shown in Figure 8.22. Note that percent covers, for all plant species combined, always sum to substantially less than 100% because from 12% to 50% of each field is covered with litter (Inouye et al. 1987a) and an additional 0% to 30% is bare mineral soil (Fig. 8.2).

Nitrogen Accumulation

In the broadest outline, the data from the experimental nitrogen gradients at Cedar Creek support the hypothesis that the successional pattern is influenced by the temporal nitrogen gradient. Early successional fields are dominated by annuals (Fig. 8.2), which tend to reach their peak abundance at low rates of nitrogen supply and decline in abundance at greater rates of nitrogen supply on the experimental gradients (Tilman 1987a). Long-lived grasses and long-lived forbs, especially native prairie species, increase in abundance during succession and along the experimental nitrogen gradients. Woody plants increase in abundance during succession and along the experimental nitrogen gradients. At this level of resolution, then, there is support for the hypothesis that the slow accumulation of nitrogen could cause of the pattern of secondary succession at Cedar Creek. However, when the data are viewed on a finer time scale and on a species-by-species basis, it is clear that other processes must be important.

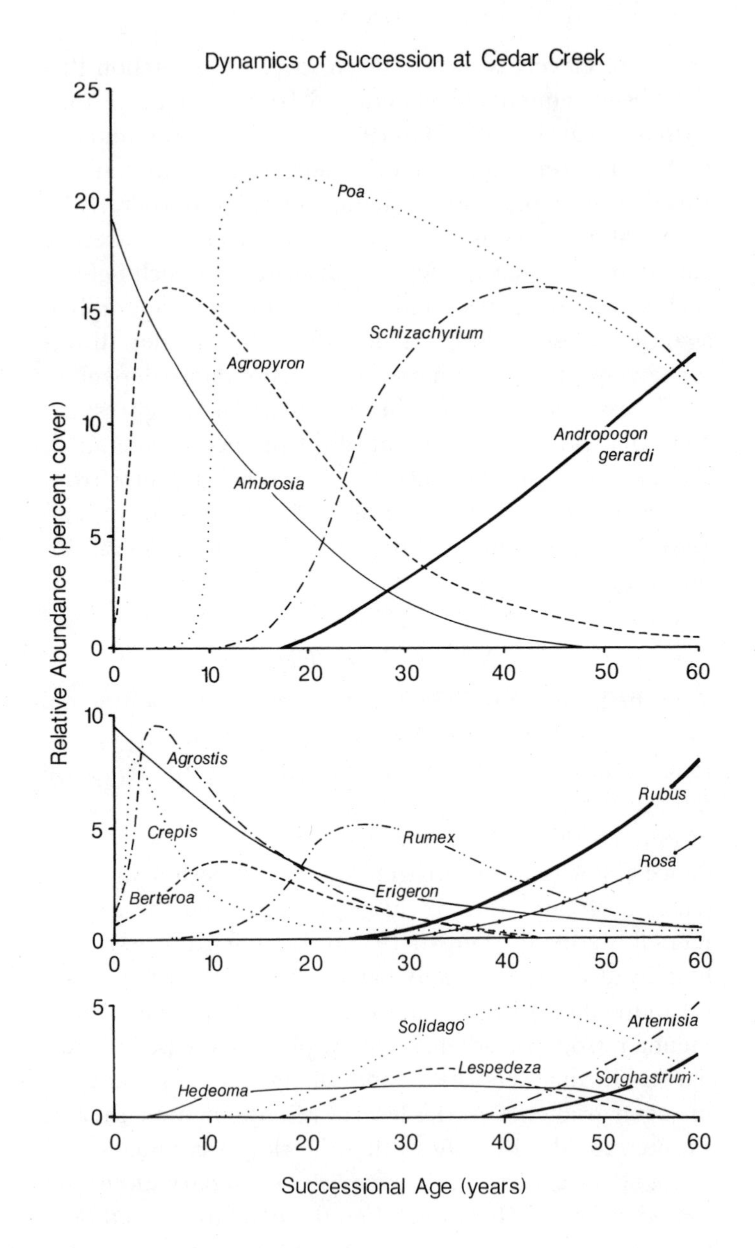

Dynamics of Succession at Cedar Creek
Relative Abundance (percent cover)
Successional Age (years)
Poa
Agropyron
Schizachyrium
Ambrosia
Andropogon gerardi
Agrostis
Crepis
Berteroa
Rumex
Erigeron
Rubus
Rosa
Solidago
Artemisia
Lespedeza
Sorghastrum
Hedeoma
0
5
10
15
20
25
30
40
50
60

Several of the common species have a temporal pattern of abundance during succession that is consistent with soil nitrogen being a major determinant of their successional dynamics. For instance, *Ambrosia artemisiifolia* declined in abundance following nitrogen fertilization, was less abundant in the more nitrogen-rich areas within fields and in the more nitrogen-rich fields, and declined in abundance during successional time (Fig. 8.22). *Erigeron canadensis* generally decreased in abundance with increasing soil nitrogen in the old field survey (Fig. 8.18) and during succession (Fig. 8.22), but did not have a consistent response to nitrogen addition. *Rubus* sp. increased in abundance with nitrogen addition, was more abundant in more nitrogen-rich areas within and among fields (Fig. 8.18), and increased in abundance during succession (Fig. 8.22). *Berteroa incana* and *Artemisia ludoviciana* had Gaussian responses to nitrogen in the old field survey (Fig. 8.18) and had successional dynamics that were consistent with these, considering the total nitrogen levels at which they reached peak abundance and the actual total soil nitrogen levels observed during succession at Cedar Creek. However, other species, such as *Agrostis scabra*, *Hedeoma hispida*, *Crepis tectorum*, *Rumex acetosella*, and *Lespedeza capitata*, decreased in abundance with nitrogen both within and among fields, and often decreased in abundance along the experimental nitrogen gradients, but had periods ranging from 3 to 30 years during which their abundance increased with time

FIGURE 8.22. Dynamics of succession at Cedar Creek, based on an old field chronosequence and observations within some permanent plots. For each of 17 common old field species, its abundance (as estimated by percent cover) is graphed against successional time. This figure was prepared using the field average percent covers for each species as determined by a survey of 22 old fields (described in Inouye et al. 1987a) and as determined by Barbara Delaney's repeated surveys of 11 successional strips at Cedar Creek. I used these 33 data points to draw a smooth curve showing the dynamics of succession for each of these 17 species. Some of the late successional species are absent from fields for periods ranging up to the first 40 years of succession.

during succession at Cedar Creek (Fig. 8.22). Thus, each of these species was absent or rare for a portion of early succession during which it would have been expected to be very abundant based on its response to nitrogen and the tendency for nitrogen to increase during succession. Similarly, *Solidago nemoralis* was absent from succession on the most nutrient-poor soils of early succession, and increased in abundance slowly through time, even though it declined in abundance with nitrogen addition. *Poa pratensis* reached its peak abundance earlier in succession than *Schizachyrium scoparium* (Fig. 8.22), even though both the nitrogen addition experiments and the dependence of the abundances of these species on soil nitrogen (Fig. 8.18) predict that *Schizachyrium* should be dominant before *Poa* if successional dynamics are determined just by the observed increase in nitrogen through time. Moreover, *Agropyron repens* reached its peak abundance in the 4th to 8th years of succession, and then declined, even though it increased in abundance dramatically and consistently with nitrogen addition for 6 of the 7 experimental nitrogen gradients.

Thus, much of the actual dynamics of succession cannot be explained as an equilibrium result of competition for gradually increasing soil nitrogen during succession. Indeed, for three of the five most abundant species during succession—*Agropyron*, *Poa*, and *Schizachyrium*—their order of peak occurrences is the exact opposite of what would be expected based on the nitrogen accumulation hypothesis (compare Fig. 8.18 with Fig. 8.22). Might these be explained by colonization rates or as transient dynamics?

Colonization

For a species to grow in a habitat, it must be present in that habitat. For a species that is a poor competitor on nutrient-poor soils, slow colonization rates would not greatly influence its successional dynamics on an initially poor soil. However, poor colonization rates, whether caused by seeds with short dispersal distances or low densities of seed sources in the region of a newly disturbed site, would have great impact on species that were good competitors for low nutrient habitats. *Ambrosia artemisiifolia*, a dominant of early succession at Cedar Creek, is a seed banking annual. It is also a common agricultural weed. Its seeds were present in these fields when abandoned, and its immediate dominance of these fields could be ascribed just to its initial abundance. However, it maintained a high abundance in these fields (especially in more nitrogen-poor areas within the fields) for more than 20 years, and declined dramatically with nitrogen addition, indicating that its initial abundance was not the only determinant of its pattern of dominance during early succession. *Erigeron canadensis*, a winter annual with seeds that are easily dispersed by wind, is another dominant of newly disturbed fields. Although it is not a seed banking species, it is an agricultural weed, and thus seeds were present in the fields when farming stopped. Its presence in these fields during the early stages of succession may be strongly related to its role as a weed and its well-dispersed seeds. It showed little consistency in its response to nitrogen addition experiments.

Based on their old field nitrogen distributions (Fig. 8.18) and their responses to nitrogen addition, *Crepis tectorum*, *Agrostis scabra*, *Rumex acetosella*, and *Lespedeza capitata* would be expected to be dominant in the nitrogen-poor soils of newly abandoned fields. However, these species required from 2 to 30 years to reach their peak abundance (Fig. 8.22). This 2 to 30 year delay may have been caused by slow colonization. This hypothesis is supported by a correspondence between likely dispersal rates of these species and the length of the delay before each reaches its peak abundance. *Agrostis* has a tumbleweed mechanism of seed dispersal and *Crepis* has small seeds that have a parachute-like plume that aids dispersal. These two species reach their peak abundance within 2 to 5 years (Fig. 8.22). In contrast, *Rumex* and

Lespedeza have large seeds with no dispersal structures, and are thus likely to be slow colonists. They require 20 to 30 years to attain their peak biomass. Further, the peak biomass each attains is low, perhaps because, by the time they have colonized a field, the field average nitrogen levels are no longer favorable to them, and they are constrained to locally nutrient-poor areas. In such areas, though, each species can be locally abundant.

Transient Dynamics

In addition to dispersal mechanisms, differences in maximal rates of vegetative growth, such as those that could be caused by different patterns of allocation to leaves, may explain some of the successional dynamics observed at Cedar Creek. For instance, *Lespedeza capitata,* a nitrogen-fixing legume, is likely to have a low maximal growth rate (Grime and Hunt 1975; Fig. 3.2). This low maximal growth rate could cause it to be initially suppressed by species that are poorer competitors for nitrogen but have higher maximal growth rates. Together with slow dispersal rates caused by large seeds, this may explain the 30 years required for it to reach its peak abundance. Conversely, high maximal growth rates could allow species that are inferior equilibrial competitors (for the loss rate and nutrient supply rate of a habitat) to have periods of transient dominance. Just this was observed in the nitrogen addition experiments. As already discussed, annual plants have higher maximal rates of vegetative growth than perennials (Chapter 3; Fig. 3.2). Annual plants increased highly significantly in abundance along the experimental nitrogen gradient in the 14-year-old field in the first year of the experiment, whereas perennials decreased highly significantly (Tilman 1987a). In the second year of the experiment, neither group responded significantly to the gradient. By the third and fourth years, annuals decreased significantly along the gradient, and perennials increased (Tilman 1987a). Thus, nitrogen addition led to a period of transient dominance by annuals, which were displaced by the more slowly growing perennials by the third and fourth years of the experiment.

A similar pattern occurs on any locally disturbed site at Cedar Creek. The first plants to dominate a local disturbance, such as the mound of bare soil created by a plains pocket gopher, are seed-banking annuals (Tilman 1983, 1987a). Within a few years, these are displaced by perennials, many of which invade the plot vegetatively from its edges. Similarly, the newly disturbed experimental nitrogen gradients were dominated by annuals the first year, but by perennials the following years. A few individuals of these annual species seem occasionally to escape suppression by perennials in high nitrogen plots, probably because they germinated in an area in which perennials were locally rare. The rapid growth rates of these annuals, combined with their plasticity in maximal size, allows them to become quite large and thus to locally suppress neighboring perennials. However, these are rare events, and perennials invade these microsites and suppress the annuals within a year or two. The most common annuals species to do this on the high nitrogen plots are *Erigeron canadensis* and *Setaria glauca*.

A variety of species have shown some dramatic patterns of transient dominance in the undisturbed nitrogen gradient experiments. Following nitrogen addition to the preexisting vegetation of the 14-year-old field, the relative abundance of *Ambrosia artemisiifolia* (ragweed, a native prairie annual) increased 12-fold on the highest nitrogen treatment during the first growing season, but *Ambrosia* had almost disappeared by the second growing season (Fig. 8.23). *Berteroa incana* responded similarly. Both of these species, and essentially all other species, were displaced from the highest nitrogen treatment by *Agropyron repens*.

In the high nitrogen plots in the 25-year-old field (Fig. 8.24), *Rumex acetosella* increased in relative abundance the first year, declined, and increased again in 1986. *Panicum*

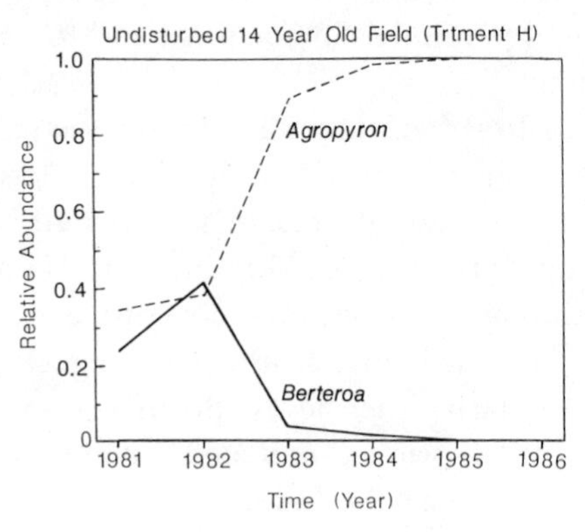
Undisturbed 14 Year Old Field (Trtment H)
Relative Abundance
1.0
0.8
0.6
0.4
0.2
0
Agropyron
Berteroa
1981
1982
1983
1984
1985
1986
Time (Year)

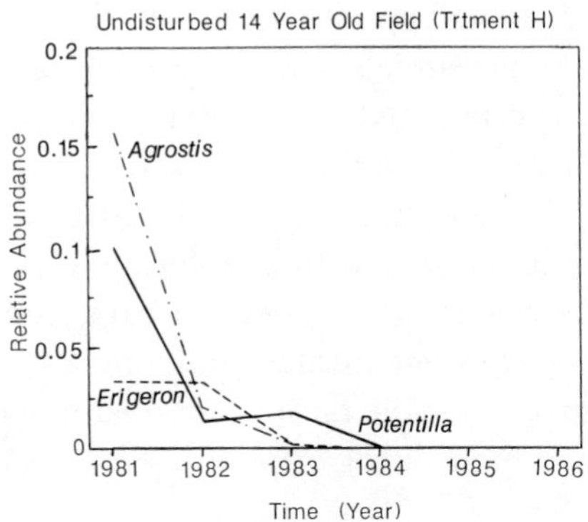
Undisturbed 14 Year Old Field (Trtment H)
Relative Abundance
0.2
0.15
0.1
0.05
0
Agrostis
Erigeron
Potentilla
1981
1982
1983
1984
1985
1986
Time (Year)

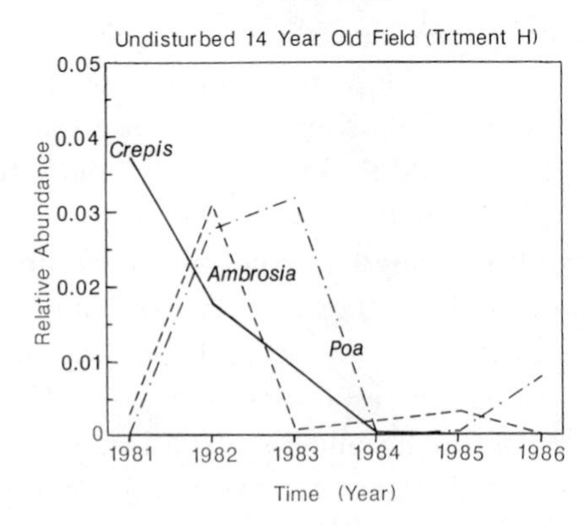
Undisturbed 14 Year Old Field (Trtment H)
Relative Abundance
0.05
0.04
0.03
0.02
0.01
0
Crepis
Ambrosia
Poa
1981
1982
1983
1984
1985
1986
Time (Year)

oligosanthes increased in abundance for three years, and then declined. *Poa pratensis* had a gradual increase in abundance for the first four years. *Agropyron repens* increased in abundance for the first five years. The relative abundance of *Schizachyrium scoparium* was unaffected by the treatment the first two years, and then declined. *Lespedeza capitata*, a legume, declined in relative abundance in the first year of nitrogen addition, and from then on.

Similar patterns of transient dominance were shown by the common species of the 48-year-old field (Fig. 8.25) and the field of native savanna (Fig. 8.26). Some species reached their peak abundance the first year, some the second, third, fourth or fifth years.

Thus, in all four fields, which ranged from early successional fields to native savanna, the immediate response to nitrogen addition was markedly different from the pattern observed three years later, and that differed from the response seen after five years. Indeed, nitrogen addition led to a series of periods of transient dominance by a variety of different species. It is quite unlikely that such transitions have stopped, even five years after the experiment began. Based on the responses of woody species such as *Rubus*, *Rosa*, and *Rhus* to nitrogen fertilization, and on the correlations we have observed in the old field survey between soil nitrogen and woody plant biomass, the ultimate "equilibrium" community in these high nitrogen plots is likely to be dominated by woody plants, especially trees. If this is so, it will be decades before these experiments reach an equilibrium, and almost all of the species dominant in the fifth year

FIGURE 8.23. The observed dynamics of secondary succession in treatment *H* plots (high nitrogen plots; see Fig. 8.9) of the undisturbed 14-yr-old field at Cedar Creek. The data for "1981" are actually the 1982 field averages for the completely unmanipulated control, treatment *I*. These are the best available estimate of what the vegetation was like before nutrient addition. The data for 1982 and beyond show the effect of nutrient addition near the end of the growing season. Only the most abundant species are shown.

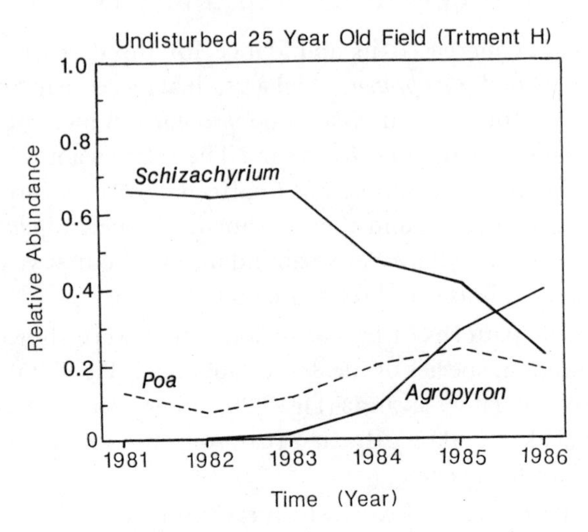

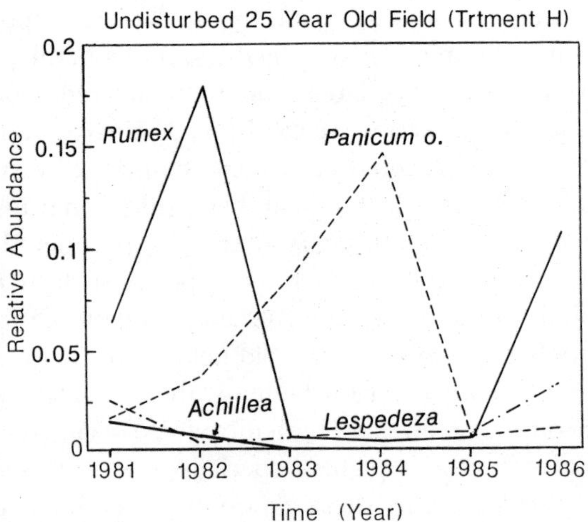

FIGURE 8.24. The observed dynamics of secondary succession in treatment *H* plots (high nitrogen plots) of the undisturbed 25-yr-old field at Cedar Creek. See legend to Figure 8.23 for more details.

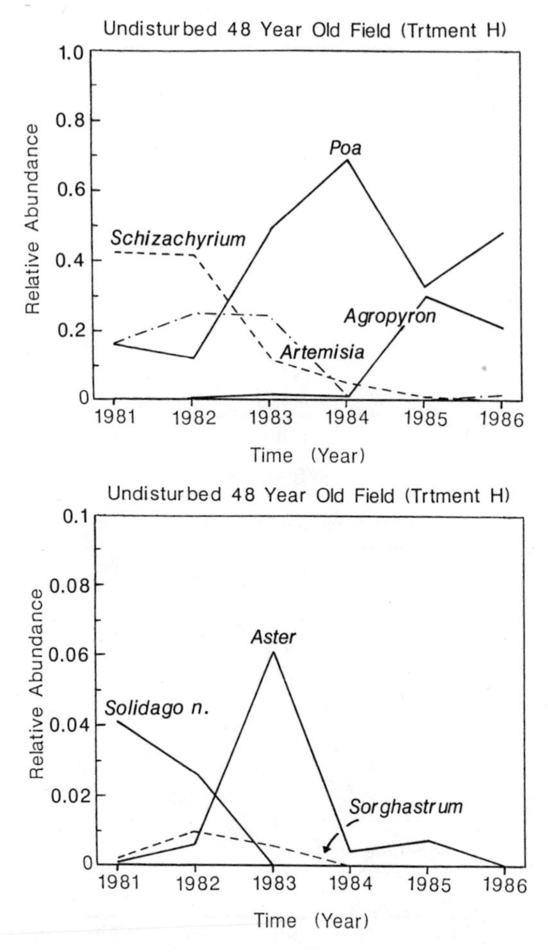

FIGURE 8.25. The observed dynamics of secondary succession in treatment *H* plots (high nitrogen plots) of the undisturbed 48-yr-old field at Cedar Creek. See legend to Figure 8.23 for more details.

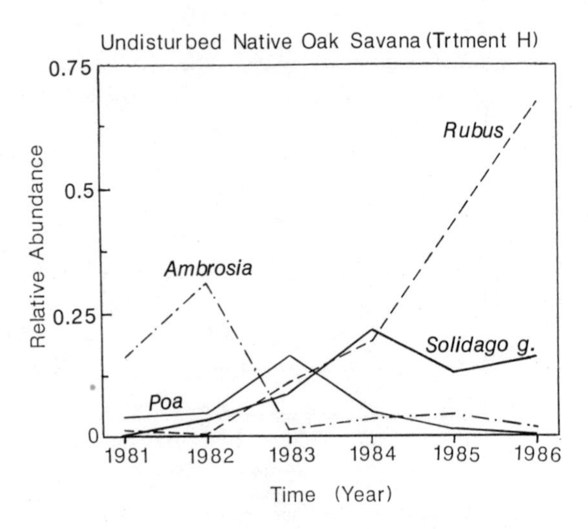

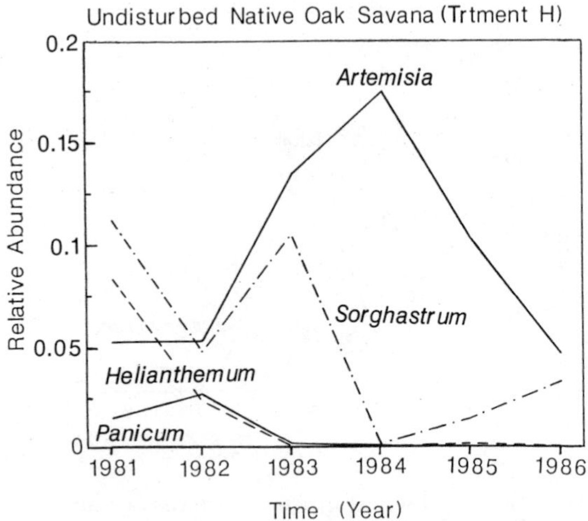

FIGURE 8.26. The observed dynamics of secondary succession in treatment *H* plots (high nitrogen plots) of the undisked native oak savanna at Cedar Creek. See legend to Figure 8.23 for more details.

of the experiment may well prove to be transients. Indeed, data collected in the sixth year show *Agropyron* being displaced from some high-nitrogen plots by *Rubus*.

These transient dynamics are qualitatively consistent with the predictions of Chapters 6 and 7. The first species to attain their peak abundance were annuals or short-lived perennials, which are likely to have high maximal rates of vegetative growth. The species dominant in the fifth year, though, have slower growth rates and were often rare or absent from the plots at the start of the experiment. In the 14-year-old field, in which *Agropyron repens* comprised about 30% of the biomass of the plots at the start of the experiment, the period of transient dominance by other, faster-growing species was a single year. In the 25- and 48-year-old fields, in which *Agropyron* was rare (< 1%) and locally absent at the start of the experiment, other species attained transient dominance for the first three years. Similarly, in the native oak savanna, in which a relatively slowly growing woody vine was dominant after five years, other species had periods of transient dominance during the first three or four years of the experiment.

I have measured the rates of vegetative growth of 9 Cedar Creek plants, as well as the dependence of their growth rates on soil nitrogen levels (Tilman 1986b). These rates provide further support for the role of transient dynamics. The initial dominant of the 14-year-old field, *Ambrosia artemisiifolia*, attained 2.5 times greater biomass than did *Agropyron repens* during 12 weeks of growth on a rich soil. Moreover, *Ambrosia* was 30 cm tall at that time, more than double the height of *Agropyron*. Thus, the more rapid growth rate of *Ambrosia* may explain its ability to overtop and suppress *Agropyron* during the first year of the nitrogen gradient experiments (Fig. 8.23). However, *Agropyron* displaced *Ambrosia* from the second year on, probably because the energy and nutrients stored in the roots and rhizomes of *Agropyron* allow it to grow more rapidly than an annual early in the the season. *Schizachyrium scoparium*, which increased in absolute abundance but not in relative abundance immediately following nitrogen addition (Tilman 1987a), attained the least biomass of all 9 of the species in the greenhouse growth experiments (Tilman 1986b). Its failure to ever have a period of transient dominance in its relative abundance following nitrogen addition is probably caused by this low growth rate. These same greenhouse data, however, also support the hypothesis that equilibrial competitive ability for nitrogen is a major determinant of the successional sequence. These data showed that there was a highly significant rank order correlation between the biomass a species attained on nitrogen-poor soils and its order of occurrence during succession (Tilman 1986b). Early successional species produced greater biomass on poor soils than later successional species.

The results observed in the disturbed (disked) plots also support the role of transient dynamics in secondary succession. For instance, in the 25-year-old field, the initial dominants of disturbed but unfertilized plots were the annuals *Setaria glauca* and *Polygonum convolvulus*, with *Polygonum* much more abundant than *Setaria*. These two were replaced by the short-lived perennials *Agrostis scabra* and *Rumex acetosella*, and then by the slowly-growing *Schizachyrium scoparium*. The high nitrogen plots had quite different dynamics, with *Polygonum* more abundant than *Setaria* the first year, *Setaria* attaining its peak relative abundance the second year, *Poa pratensis* and *Panicum oligosanthes* reaching their peak relative abundances the third year, and *Agropyron* being dominant by the fourth and fifth years. On both poor soils and rich soils, the pattern of species replacements following a disturbance represents transitions from fast to slow growing species (Tilman 1986b; Grime and Hunt 1975). Moreover, as is predicted by the theory summarized in Figure 7.5, the trajectory of dominant species is quite different for secondary successions on poor soils than for those on rich soils.

CAUSES OF SUCCESSION

This study of succession at Cedar Creek has shown that the dynamics of secondary succession on a poor soil are influenced by the pattern of nutrient accumulation during succession, by differentiation in competitive abilities along a nutrient gradient, by differences in maximal growth rates of species, and possibly by the dynamics of colonization. Each of these processes can explain some aspects of the successional dynamics at Cedar Creek, but each cannot explain other aspects.

Even on the nutrient-poor soils of Cedar Creek, much of the dynamics of succession seem to be transient dynamics. The sequence of dominance first by *Ambrosia*, then *Agropyron*, *Poa*, *Schizachyrium*, and finally *Andropogon* (Fig. 8.22) is inconsistent with both the observed responses of the three intermediate species to nitrogen addition and their distributions with respect to total soil nitrogen in individual fields. The early dominance by the rapidly-growing *Agropyron*, followed by the less rapidly-growing *Poa*, and then by the slow-growing *Schizachyrium* (Fig. 8.22) is consistent with observed differences in their maximal growth rates (Tilman 1986b; Cowan 1986), and thus may be an example of transient dynamics. If this is correct, transient dynamics may be an important determinant of succession at Cedar Creek for the first 30 or 40 years (Fig. 8.22). The importance of transient dynamics during secondary succession at Cedar Creek is bolstered by root:shoot dynamics during succession. As illustrated in Figure 7.5A and B, the period of transient dynamics on poor soils is predicted to be a period during which species with low root:shoot ratios are displaced by species with increasingly larger root:shoot ratios. Measured root:shoot ratios of the mixed vegetation of Cedar Creek fields show just this pattern. Root:shoot ratios in a one-yr-old field are about 1:1. They are about 5.2:1 in a 15-yr-old field, and 11:1 in a 50-yr-old field (R. McKane, pers. comm.). This supports the hypothesis that species like *Agropyron* attain dominance during early succession because of higher maximal growth rates caused by lower root:shoot ratios, and that these are displaced by species such as *Schizachyrium* that are slower growing but are superior long-term competitors for soil nitrogen levels. This same pattern holds when root:shoot ratios of individual species are considered. When grown in monocultures on different soil types in the field, the five most abundant successional grasses at Cedar Creek have root:shoot ratios on poor soils that agree perfectly with the hypothesis that their succession sequence is determined by transient dynamics (Table 8.1; Wedin and Tilman, unpublished data). *Agrostis scabra*, which is the first of these species to attain dominance during succession (Fig. 8.22), has the lowest root:shoot ratio of these five species. *Andropogon gerardi*, the latest successional grass, has the highest root:shoot ratio (Table 8.1). Moreover, the data collected in the second year of an experiment in which 8 Cedar Creek herbs were grown in soil mixtures in 18-liter pots in the field showed a similar pattern (Tilman and Cowan, unpublished data). There was a significant rank order correlation between the observed root:shoot ratio of each species and its time of peak abundance during succession ($r_s = 0.81$, $n = 8$, $P < 0.05$). These various pieces of evidence support the hypothesis that much of the successional sequence at Cedar Creek may be the transient dynamics of competitive displacement.

SUMMARY

Upon abandonment from agriculture, the old fields of Cedar Creek Natural History Area, Minnesota, undergo successional dynamics that seem to be determined by (1) the abilities of the species to compete for nitrogen and light, (2) the slow rate of nitrogen accumulation on these nitrogen-poor soils, and (3) the allocation patterns and thus the maximal growth rates of the species. Colonization rates may also

TABLE 8.1. Root:shoot ratios for the five major grasses of Cedar Creek secondary succession were determined by growing each species in replicated monocultures along a nitrogen gradient. Root:shoot ratios at low soil nitrogen are the y-intercepts of regressions of root:shoot ratios (y) against total soil nitrogen (x), based on 18 one-year-old monoculture plots for each species. Successional ranks are from Figure 8.22. Root:shoot data are unpublished data of D. Wedin and D. Tilman. Spearman rank order correlation between successional rank and root:shoot ratio is $r_s = 1.0$; $N = 5$; $P < 0.05$. Based on 1986 data.

Species	Successional Rank	Root:Shoot at Low Nitrogen
Agrostis scabra	1	0.42:1
Agropyron repens	2	1.0:1
Poa pratensis	3	1.2:1
Schizachyrium scoparium	4	1.5:1
Andropogon gerardi	5	2.3:1

play a role. These same factors, but especially competitive ability for nitrogen and light, influence the local abundances of species within any given field. The species that are dominant during the early stages of secondary succession at Cedar Creek, such as *Ambrosia artemisiifolia*, *Agrostis scabra*, and *Agropyron repens*, have higher maximal rates of vegetative growth, lower root:shoot ratios, and can grow more rapidly on nitrogen-poor soils than such late successional species as *Andropogon gerardi*, *Rubus* sp., *Sorghastrum nutans*, and *Artemisia ludoviciana*. Thus secondary succession on a poor soil is a multi-causal process.

The results of this study suggest that secondary successions on poor soils are intermediate between primary successions and secondary succession on rich soils. Although all successions are likely to be multi-causal, the relative importance of different processes may change along the continuum from primary successions to secondary successions on rich soils. Primary successions are highly influenced by the process of nutrient accumulation in the soil and the subsequent changes in light penetration to the soil surface and vertical light profiles. Because soil development is a relatively slow process and because the initial substrates of primary successions are extremely nutrient limited, colonization rates and the transient dynamics of competitive displacement are unlikely to be of great importance during primary successions. At the other extreme, secondary successions on rich soils are unlikely to be influenced by long-term soil dynamics, but should be highly dependent on the colonization rates of various species and on differences in maximal growth rates (i.e., transient dynamics). Secondary succession at Cedar Creek is an intermediate case. The nitrogen-poor soils of Cedar Creek and the slow rate of accumulation of nitrogen seem to be important factors allowing the long-term persistence of "early successional" species and delaying the re-establishment of dominance by late successional species, especially woody species. However, many species have periods of peak dominance that cannot be explained by the nitrogen accumulation hypothesis, but are consistent with the hypothesis that transient dynamics or colonization influence succession. This suggests that the initial nutrient status of the soil is an important factor controlling the relative importance of colonization, transient dynamics, and nutrient accumulation in successions.

CHAPTER NINE

Questions and Conclusions

The preceding chapters raise a number of questions that I will consider before summarizing this book.

PHENOTYPIC PLASTICITY

As discussed in Chapter 2, morphological plasticity can be of adaptive value. For instance, a plant limited by a soil nutrient could increase its growth rate by decreasing its leaf or stem allocation and thus increasing its root biomass and rate of nutrient uptake. Similarly, a light-limited plant that increased its leaf area and/or its height would obtain more light and have a more rapid growth rate. Physiological mechanisms can also be important. A phosphorus-limited plant that increased the number of uptake sites on the cell surfaces of its roots could obtain more phosphorus and grow more rapidly. A light-limited plant that increased its photosynthetic efficiency, such as by increasing leaf chlorophyll and protein content, could grow more rapidly. Such morphological and physiological adjustments are favored even if they affect the ability of a plant to use some other resource, as long as that other resource does not currently limit the plant's growth (Iwasa and Roughgarden 1984; Bloom, Chapin, and Mooney 1985).

The "foraging behavior" of a plant comes from its ability to adjust its physiology and morphology. Optimal foraging theory for essential resources predicts that a plant should allocate its potential growth and adjust its physiology so as to be equally limited by all essential resources (Rapport 1971; Covich 1972; Tilman 1982; Iwasa and Roughgarden

1984; Bloom, Chapin, and Mooney 1985). Bloom, Chapin, and Mooney (1985) reviewed much of the literature on plant physiological plasticity, and found that plants do seem to adjust their physiology in the direction that would cause them to be more equally limited by several resources. However, nutrient addition experiments performed in a variety of natural vegetation stands have frequently shown that most species are limited by a single nutrient, not jointly limited by several nutrients (Lawes and Gilbert 1880; Milton 1947; Brenchley and Warington 1958; Willis 1963; Tilman 1980, 1982, 1983, 1987b). The available evidence, though, is not a good test of the foraging theory prediction. To test the equal limitation hypothesis fully will require the addition of all possible combinations of nutrients added two at a time, three at a time, etc. Plasticity has not been explicitly included in most of the theory developed in this book. Are the conclusions that have been drawn here so far validly applicable to phenotypically plastic plants? If plants are phenotypically plastic, why can't a single species take on the morphology and physiology that is optimal for all habitats? Could plasticity so complicate the process of interspecific competition that no simple theories will ever be able to describe plant interactions?

To explore the effects of morphological plasticity on the growth of a single species and on interspecific competition, I modified the model ALLOCATE to have the pattern of allocation for each plant at any given point in time be influenced by its current pattern of resource limitation. If a plant were currently limited by the soil nutrient, it would increase its proportional allocation to roots and decrease its proportional allocation to leaves and stems. If a plant were limited by light, it would increase its proportional allocation to either leaves or stem, whichever led to the higher net growth rate under the current conditions, and decrease its proportional allocation to roots. As such, it would forage for the soil resource and light in a more optimal manner than individuals with a fixed allocation pattern.

What happens when physiologically identical plants are morphologically plastic? If a single population grows in a particular habitat (i.e., at a fixed rate of nutrient supply and a fixed loss rate), average allocation patterns are size (height) dependent (Fig. 9.1). Allocation also depends on the nutrient supply rate of the habitat. In a nutrient-poor habitat (TN = 50), allocation to leaves ranged from 45% to 75%, and allocation to roots ranged from 10% to 55%. Although this is considerable height-dependent variation, these ranges barely overlap the ranges of height-dependent variation for the same species growing on a richer soil (TN = 200). For a given, fixed physiology, more fertile soils favor lower allocation to roots and higher allocation to leaves or stems.

What would happen if two morphologically plastic plants were to compete? Clearly, if the plants had the same underlying physiology and both adjusted their morphology in the same manner, they would be functionally identical. What would happen though, if they had different physiologies? Let one species have a lower requirement for a soil nutrient, because it requires less nutrient to produce a unit of new tissue and less per unit tissue to attain a given photosynthetic rate. This is species *A*, whose variable allocation responses were presented in Figure 9.1. Let the other species, species *B*, have a lower requirement for light, i.e., a higher photosynthetic rate at low light levels. Both species have identical nutrient- and light-saturated photosynthetic rates, identical seed sizes, identical heights at maturity, identical respiration rates, and an identical allometric relation between stem mass and height. When these species competed, species *A* was most abundant in nutrient-poor habitats and species *B* was most abundant in nutrient-rich habitats (Fig. 9.2), just as would have happened without plasticity.

Interspecific competition compressed the range of morphological variation in both species. Compared to its morphology in the absence of interspecific competition, inter-

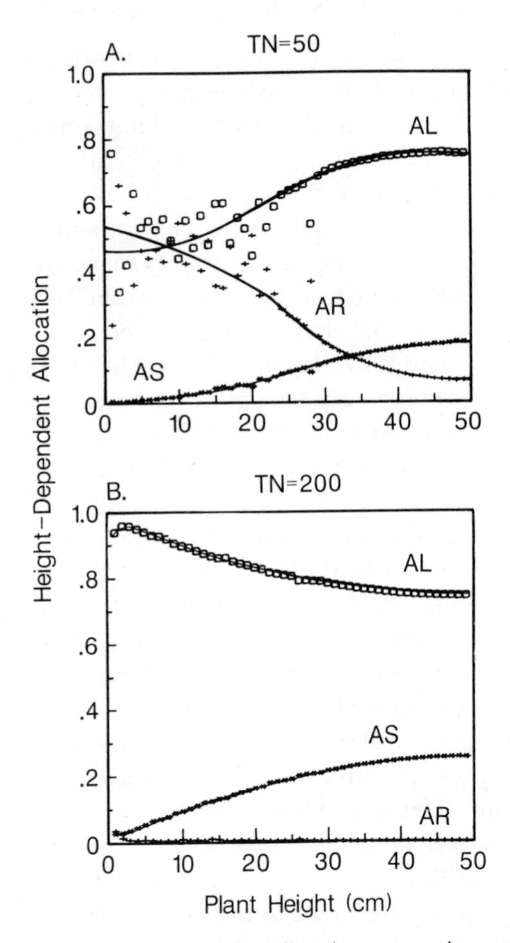

FIGURE 9.1. (A) Height-dependent allocation on a nutrient-poor soil (TN = 50) in a species in which individual plants are morphologically plastic, modifying their allocation in response to resource availabilities. The height-dependent allocations to leaves (AL), to roots (AR), and to stems (AS) are long-term average allocation patterns. (B) Allocation patterns of this same species on a richer soil (TN = 200). Note that, in comparison with their allocation patterns on the poor soil, these individuals allocate much more to leaves, somewhat more to stems, and much less to roots. Results are based on a modified version of ALLOCATE in which individuals increased their proportional allocation to the structures that would lead to the greatest increase in their current rate of biomass accrual.

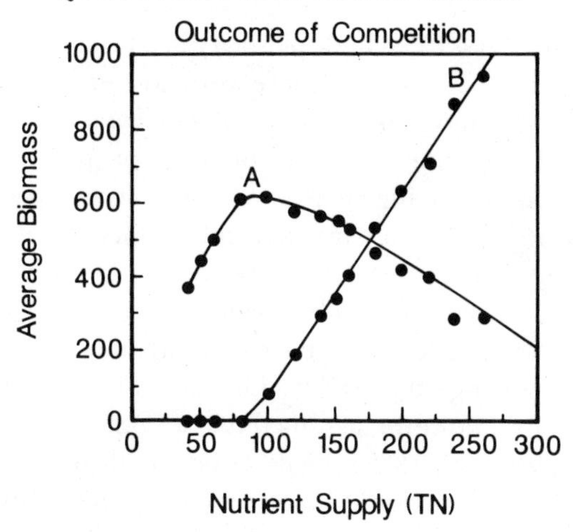

FIGURE 9.2. Results of simulations of competition among these two morphologically variable species. Species *A* displaced species *B* in nutrient-poor habitats and both species coexisted in more nutrient-rich habitats. Species *B* displaced species *A* from even richer habitats (TN > 700; not shown).

specific competition caused each species to have allocation patterns that reinforced the underlying physiological differences between the species. For instance, species *A*, which is a better competitor for nutrients, was forced by interspecific competition to shift its allocation back in the direction of higher root allocation. The greater the number of competing species, the more similar the species must be, and the more each might be limited by interspecific competition, at equilibrium, to a more narrow range of habitat types and thus allocation patterns.

Although morphological plasticity can influence competition along a nutrient supply gradient, there are several reasons why it is unlikely to have a major effect on the conclusions reached in the preceding chapters. First, even though plasticity can allow the persistence of species and modify the relative abundance of species along nutrient gradients, it does not seem to change the qualitative pattern of species distributions along these gradients, if all species are equally plastic. Second, plasticity itself may have costs. The model above assumed there were no costs. If there are costs to continuous monitoring of environmental conditions and adjustments in morphology or physiology in response to a changing environment, then such costs would decrease the benefit of plasticity, and limit its extent. Third, the greater the number of competing species, the smaller should be the range of viable morphologies within each species. Fourth, all species have structurally-imposed limits to their morphological plasticity. The model of plasticity presented above, however, assumed that a single genotype was sufficiently plastic that it could produce all possible morphologies. Thus, a single genotype was assumed capable of becoming a single-celled soil alga, a moss-like plant, an herb, a woody vine, and a tree.

Although plant species are morphologically and physiologically plastic, the range of such plasticity within an individual or within a given species is minuscule compared to the range of variation in the morphology and physiology of different species. Soil algae never attain the stature or form of mosses. Grasses never attain the form of shrubs and shrubs never become trees. Early successional annuals such as *Ambrosia artemisiifolia* and *Chenopodium album* can have their height at the end of the growing season differ by a factor of 5 depending on the habitat in which they grew. However, the structural materials of which they are composed, and the manner in which those structural materials are arranged, are incapable of supporting individual plants that are much taller than the maximum heights observed. A 5 m tall *Ambrosia artemisiifolia* is a structural impossibility. It would buckle under its own mass. Root:shoot ratios are also phenotypically plastic, but never differed by more than a factor of 2.5 within each of 8 Cedar Creek species grown on a nitrogen gradient with a 15-fold range in nitrogen availability (Cowan 1986; Cowan and Tilman, in prep.). Similarly, Chapin (1980, p. 236) reported that "a 100-fold drop in availability of a limiting nutrient causes a 1.5- to 12-fold increase in root:shoot ratio," and that "the combined compensatory effects of increased root:shoot ratio and increased root absorption capacity do not fully compensate for reduced nutrient availability." Thus, there are limits to morphological and physiological plasticity. These limits constrain a plant to being a superior competitor for a relatively narrow range of habitat types.

A 1 m tall oak tree is clearly not a structural impossibility, but if 1 m were the maximal height at maturity for oaks in a habitat, individuals that produced the usual woody stems of oak seedlings would be outcompeted by individuals that produced a less expensive stem. The much more highly lignified and denser stems of woody plants, even as seedlings or saplings, are a prerequisite if the plant is ever to attain great height, for such strong structural tissues must occur throughout the entire stem of the plant for it to support itself at maturity. It seems quite unlikely that a tree could produce a thin, pithy, turgor-pressure supported stem for its growth up to a height of 1 or 2 m, and then replace this stem with dense, highly lignified material as its height increased. Thus, the properties of the materials used to construct stems and branches should differ depending on the maximum size that individuals are likely to attain. Plants that are short at maturity and short-lived can attain their maximal height using tissues and methods (such as turgor pressure) that are not suitable for plants that are taller at maturity.

The existing terrestrial vascular plants were derived, some 400 million years ago, from single-celled algae. As soils became more nutrient rich, and algal mats became thicker, "taller" individuals were favored. It is possible, for instance, that some of the early soil algae had a morphology with a few cells in the soil and a few cells projecting above the soil surface. However, diffusion of nutrients and photosynthate between the above-ground cells and below-ground cells would have been too slow to allow such plants to be much more than a few cells tall. Thus, significant increases in height required both structural materials and a system to transport nutrients and photosynthate between roots and leaves, i.e., vascularization. However, vascular systems that function adequately for short plants may not be sufficient for taller plants. There are a wide variety of vascular systems that could allow a plant to attain a height of 10 cm, few of which would be adequate for a plant 40 m tall. Each of these systems of support and transport has its own costs. For any given population, natural selection should have favored the support and transport system that led to the greatest net reproductive rate, within the constraints of the costs of producing each system and the benefits associated with each system (such as lower mortality from lodging and increased growth from efficient transport).

A major factor determining this optimum for a given population would have been the maximal height at maturity of the individuals, for this height determines the compressive load on the stem and the tendency for the stem to buckle under its own weight. In general, for plants that are relatively short at maturity, less expensive stems would have been favored, for these stems do not need to serve as the basis for the support and transport system of a tall individual. The stems would be less expensive not in terms of the cost per gram of dry weight of stem material, but in terms of the cost per meter of stem height. The added cost per meter of height for structural materials that are suitable for tall plants may have been an important limit on the range of morphological plasticity in terrestrial plants. Each individual, with the repertoire of structural materials it can produce, is limited to being a superior competitor for a relatively small range of maximal heights. At heights less than its optimum, it would be outcompeted by individuals that produced cheaper stems that were suitable for that height.

At greater heights, it would bear the costs of increased lodging or of inefficient transport.

Thus, although morphologically plastic individuals should be favored over those that have a fixed morphology, and although physiologically plastic individuals should be favored over those that are fixed, the potential range for such plasticity is small compared to the range of possible morphologies and physiologies. This means that it is unlikely that a single genotype could evolve that would be so phenotypically plastic that it could displace all other genotypes from the full range of habitat types. Further, even when plasticity is included, the qualitative outcomes of competition need not be changed. Thus, whether or not plants are plastic, different morphologies and physiologies would be favored at different points along soil-resource:light and loss rate gradients. It seems unlikely that any of the major conclusions presented in the preceding chapters would be affected in any meaningful, qualitative manner by plasticity.

There are two main advantages to plasticity. First, it allows individuals to increase their competitive ability in habitats in which they would be inferior competitors if they had a fixed morphology, i.e., in habitats that are away from the corner in their resource-dependent growth isocline. It is worth remembering (see Fig. 2.5 and Fig. 9.2) that plasticity does little to increase the competitive ability of a plant in habitats in which which it is likely to have its greatest competitive ability, which are those habitats that have resource availabilities at the corner of its isocline. If plant habitats were at equilibrium and were spatially homogeneous locally, each plant species would be restricted to those areas for which it was a superior competitor, i.e., to habitats with resource availabilities near the corner of its isocline. Because plant habitats are quite heterogeneous through space and time, there is considerable variance in the habitats an individual plant could experience. If a seed fell into a suboptimal habitat, plasticity would allow it to make the best of a bad situation. The second advantage of plasticity is that it can allow an individual to exploit variance away from the long-term mean resource levels of its habitat. Plasticity would allow a plant in a habitat with temporarily high levels of all resources to adjust its morphology so as to maximize its growth rate, and thus increase its fitness. Plasticity thus could also allow a plant to make the most of a good situation. The range in plasticity within a species may thus be a measure of the variance in resource availabilities that plants in that species have experienced during their evolutionary history.

The effect of morphological plasticity may provide an insight into the evolution of weeds. As discussed in Chapter 7, many species that are present-day agricultural weeds or are common during early stages of secondary succession seem to have evolved in nutrient-poor habitats. Such species should have lower physiological requirements for the limiting soil nutrient but higher physiological requirements for light than those that evolved on richer soils. As such, if they were to ever occur on a rich soil in an open, high light environment, they could adjust their morphology so as to have fewer roots but more leaves and stems than plants that had higher nutrient requirements. This high allocation to leaves on high nutrient soils would give them a greater maximal growth rate on such soils than species that were better competitors for light but poorer competitors for nutrients. Such a growth rate advantage would have "preadapted" such species of nutrient-poor soils for exploiting high nutrient and high light habitats, such as those that occur after a disturbance.

There is much to be learned about morphological and physiological plasticity and the effects of such plasticity on intraspecific and interspecific competition. This discussion has suggested that the presence of other species might act to limit the range of viable physiological and morphological plasticity within individuals of a species. This discussion, however, is speculative. We need more empirical data on the range and environmental correlates of plant plasticity, more experimental studies of such plasticity both within and among species, and a much firmer theoretical basis before we will be able to determine the effects of morphological and physiological plasticity on the structure and dynamics of plant communities.

SPATIAL STRUCTURE

The models presented in this book have assumed that habitats have horizontal homogeneity, with all plants in a habitat experiencing the same availabilities of soil resources and the same vertical light gradient. Clearly, real habitats are spatially patchy. Such spatial patchiness could be caused by underlying differences in soil parent material and resulting differences in soil nutrient richness, by the discrete nature of individual plants (Tilman 1982; Pacala and Silander 1985; Pacala 1986) through their effects on local nutrient availability and local light penetration, and by the discrete nature of local disturbances. Let us consider what effects such heterogeneity might have on plant community structure.

Point-to-point spatial heterogeneity in soil nutrient richness would favor different plant species at different points in space. Through time, each would become closely associated with the soil type for which it was a superior competitor (given that climate and disturbance were fixed). However, if such a model were spatially explicit, each microhabitat could include within it not only the one or two species that were superior competitors for those conditions, but also other species that were transients (in the island biogeographic sense). These transient species would most likely be species that were superior competitors for nearby, but different, soils. They would be able to persist in the habitat because of an equilibrium between immigration rates and rates of competitive displacement (Ricklefs 1987). If, as suggested in Chapter 6, competitive displacement is a slow process, it seems likely that a given local, fairly homogeneous habitat could contain many more species than there were limiting resources, with the rare species being transients maintained because their immigration rates balanced their rates of local extinction. The long-term persistence of such transient species within the full landscape, though, would be caused by the existence of habitat conditions for which each was a superior competitor.

The second view that could be taken would be that the underlying habitat was spatially homogeneous, but that the discrete nature of plants and thus of plant light interception and soil nutrient uptake, imposed point-to-point spatial heterogeneity in resource availability. This view says that habitat heterogeneity is a direct and unavoidable result of the discrete nature of individual plants. Each plant exists at a particular point or region of space and time. During its lifetime, it influences just that region, and thus only affects other plants within that region. The discrete nature of plants necessarily makes a habitat spatially heterogeneous. Such spatial stochasticity (Tilman 1982) is directly analogous to the demographic stochasticity which May (1973) discussed for low-density populations. It is an unavoidable result of the discrete nature of plants and the low numbers of individuals living in any small unit of space. The death of a single individual would create a local spot with higher light and nutrient availabilities. Such "light gaps" are an inevitable result of the discrete nature of individual plants. This means that all habitats will always have in them local sites that have temporarily higher light and higher nutrient availability than the habitat mean. This would allow plants that are better at growing at higher than average availabilities of both light and the soil resource (i.e., plants that were better competitors, at equilibrium, in habitats with higher loss rates) to persist in this habitat with species that are better at competing for the average habitat conditions. Any given habitat would then be seen as a mosaic of microsites that had different successional ages, ranging from newly open microsites caused by the death of one or a few individuals to old, closed microsites. An inescapable outcome of the discrete nature of individual plants and the spatially fixed location of each individual plant is the generation of spatial heterogeneity in resource availabilities that could allow more species to coexist than there were limiting resources.

Any given natural habitat is likely to have soil heterogeneity that was not directly caused by plants as well as spatial stochasticity caused by the discrete nature of individual plants. Both of these processes will be important in explaining the species diversity and vegetational structure of the habitat, as would an island biogeographic process of microsite colonization and extinction. Point-to-point soil heterogeneity causes different microsites to fall at different points along a gradient from low nutrient but high light habitats to high nutrient but low light habitats. Spatial stochasticity causes a habitat to contain a loss gradient, with microsites falling along a gradient from recently disturbed areas with high nutrient and high light availability to old growth areas with low nutrient and low light availability. The nutrient supply rate and loss rate gradients that are expected within any habitat would allow the persistence of many more species than there were limiting resources.

This brief discussion suggests that there may be two different ways to incorporate the effects of the "neighborhood competition" concept into models. The simplest approach may be to assume that the discrete nature of individual plants causes plant-to-plant spatial heterogeneity in resource availabilities and loss rates. Such heterogeneity could be used, in conjunction with an equilibrium model of plant competition, to map out the ranges of species traits that should be able to persist in any given habitat. Alternatively, an explicitly spatial model could be constructed in which each plant is given a location (*x* and *y* coordinates), and the resource availabilities it experiences are influenced by the locations, sizes, and consumption rates of all plants in its neighborhood. The second approach will be numerically cumbersome, but could give significant insights. Until such models are explored, the ideas above must be considered speculative.

CORRELATIONS BETWEEN DISTURBANCE AND NUTRIENT SUPPLY RATES

I have treated the loss rate of a habitat and the nutrient supply rate of its soil as if these were separate and unrelated processes. I have done this to allow an easier comparison of the theory developed in this book with patterns in natural and experimentally manipulated plant communities. Although there need not be any correlation between the loss rate of a habitat and the richness of its soil, it is likely that loss rates and soil nutrient supply rates are linked in both natural and managed communities. To the extent that there is a consistent pattern of interdependence, from habitat to habitat, such linkages could be an additional cause of patterns in natural vegetation worldwide.

Oksanen et al. (1981) hypothesized that the loss rates per unit plant biomass that are caused by herbivores should be low in very unproductive habitats because there would be insufficient plant biomass to support persistent herbivore populations. Moderately productive habitats, they suggested, should have the highest rates of loss per unit plant biomass because these could support herbivores, but would have sufficiently low herbivore biomass that predators would be rare. Highly productive habitats, however, would have sufficiently high herbivore biomass to support persistent predator populations. The predators, by feeding on the herbivores, would keep the herbivores at a lower density than they would attain in the absence of predation, and thus lead to a lower loss rate per unit biomass for plants in productive habitats. The qualitative relationship between the specific loss rate experienced by plants (loss rate per unit biomass) and the productivity of a habitat hypothesized by Oksanen et al. (1981) is illustrated in Figure 9.3A. If the trophic structure of natural communities is as they hypothesized, then this would lead to quite a different relation between productivity and vegetation structure in natural vegetation than if loss rates were unrelated to productivity. With the dependence of Oksanen et al. (1981), plant morphologies should depend on productivity as illustrated in Figure 9.3B. This would mean that the plants dominant on moderately nutrient-rich soils should have much higher maximal rates of vegetation growth than those dominant on either poor soils or rich soils (Fig. 9.3C). Further, they would be expected to have quite different morphologies than those predicted on the basis of the assumption that loss rates were constant and independent of productivity.

Plant loss rates and soil resource supply rates may be interdependent for several other reasons. For instance, fire frequency may depend on productivity. Whitney (1986) used early survey records to study the relations between soils, forest types, and fire or treefall disturbances in the presettlement forests of Roscommon and Crawford counties of Michigan. He found that the poorest soils, those formed on the coarse outwash sands of ridgetops, were dominated by jack and red pine forests and that these had the highest fire frequencies, with an 80 year return time for fire. The more productive white pine and oak forests that occurred on soils derived from medium sands had a lower fire frequency, with an estimated return time of 170 years. The most productive stands, the white pine and northern hardwood (beech and sugar maple) forests, occurred on soils derived from fine to medium fine glacial till. These had

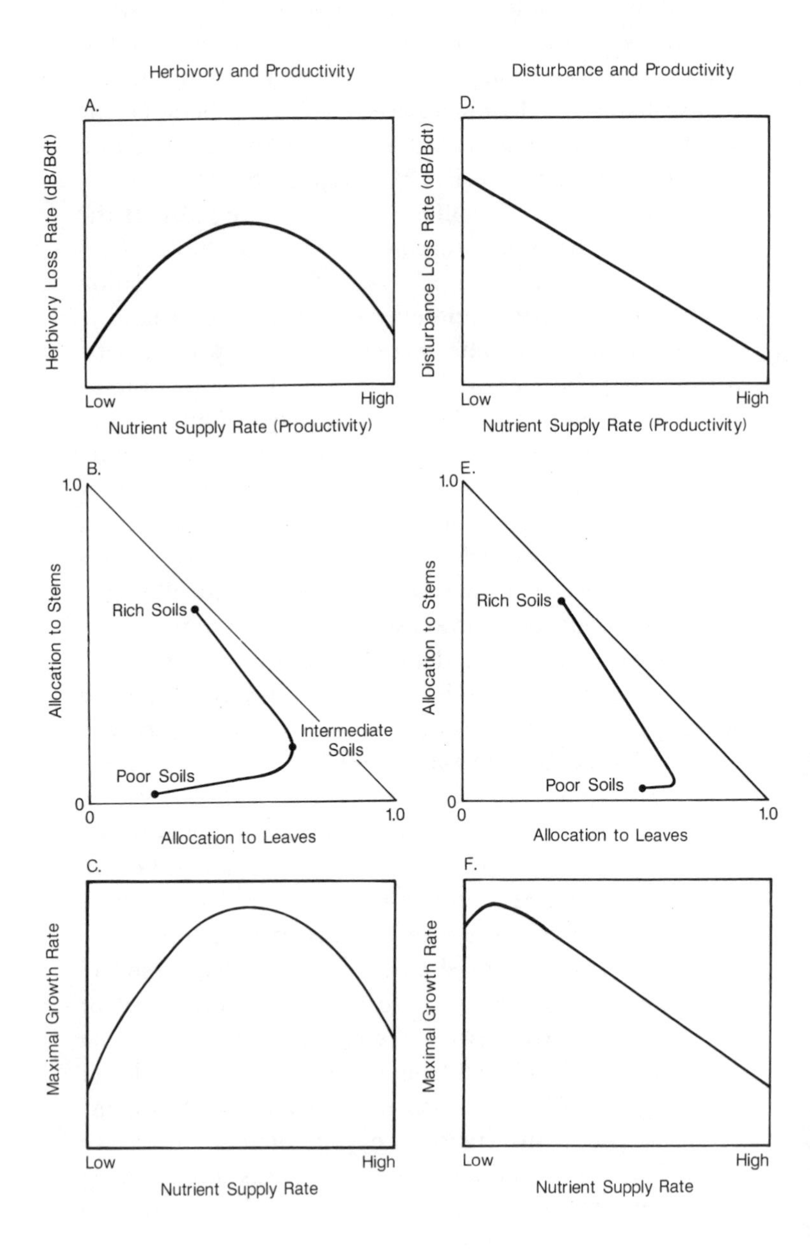

Herbivory and Productivity
Disturbance and Productivity
A.
Herbivory Loss Rate (dB/Bdt)
Low
High
Nutrient Supply Rate (Productivity)
D.
Disturbance Loss Rate (dB/Bdt)
Low
High
Nutrient Supply Rate (Productivity)
B.
1.0
Allocation to Stems
Rich Soils
Intermediate
Soils
Poor Soils
0
0
1.0
Allocation to Leaves
E.
1.0
Allocation to Stems
Rich Soils
Poor Soils
0
0
1.0
Allocation to Leaves
C.
Maximal Growth Rate
Low
High
Nutrient Supply Rate
F.
Maximal Growth Rate
Low
High
Nutrient Supply Rate

an even lower fire frequency, with an estimated return time of 1400 years. Thus, within this region, Whitney (1986) found a negative correlation between productivity and fire disturbance rates, with fire frequency correlated with the initial parent material on which the soil formed. In contrast, Whitney found that disturbances caused by tree falls, but not by fire-associated tree falls, were positively correlated with productivity, probably because taller trees with their relatively smaller root system were more susceptible to high winds. However, the treefall disturbance rates were much less than the fire disturbance rates. When both sources of loss were combined, loss rate was negatively correlated with productivity.

Such negative correlations between loss and productivity might be expected in a variety of other habitats for a variety of other reasons. For instance, slopes with exposures that give them higher disturbance rates are likely to develop poorer soils than less exposed slopes, because of the erosional and leaching losses of soil nutrients associated with disturbances (Likens et al. 1977; Bormann and Likens 1979). Even on level ground, areas with higher disturbance rates should develop poorer soils because of higher

FIGURE 9.3. (A) The relationship between the productivity of a habitat (its nutrient supply rate) and the loss rate (on a per unit biomass basis) caused by herbivores as hypothesized by Oksanen et al. (1981). (B) If loss rates and soil nutrient supply rates were interdependent as shown in part A of this figure, then plant morphologies would depend on soil nutrient richness as shown, with short, rooty plants favored on poor soils, somewhat taller, much leafier plants favored on soils of intermediate fertility, and tall plants with high allocation to stem favored on rich soils. (C) The allocation patterns favored on these soils would cause the species dominant on soils of intermediate nutrient richness to have much higher maximal growth rates than those dominant on either poor soils or rich soils. (D) An alternative hypothesized interdependence of disturbance or loss rate and nutrient supply rate. (E) Effect of the relationship of part D on the allocation patterns that lead to superior competitive ability in particular habitats. (F) Effect of these allocation patterns on maximal growth rates along a nutrient supply rate gradient. These are based on simulations using ALLOCATE.

nutrient leaching losses or wind erosion from disturbed sites. Other causes of frequent disturbances, such as landslides and major erosional events on steep slopes, would also cause these areas to have poorer soils. Thus, if all else were equal, areas with low disturbance rates would tend to develop richer soils than areas with higher disturbance rates. Wilson and Keddy (1986) found that sheltered bays in Axe Lake, Ontario, had lower loss rates for rooted macrophytes and had more organically rich substrate, whereas shores exposed to full wave action had higher loss rates and a much sandier substrate. Thus disturbance and substrate nutrient richness were negatively correlated.

For habitats in which loss rates and soil nutrient supply rates are negatively correlated, the distribution of species (morphologies) along a soil nutrient supply rate gradient (Fig. 9.3D–F) would be quite different from that described either for the case in which loss rates and soil resource supply were independent (Fig. 4.8) or the case of Oksanen et al. (1981) (shown in Figure 9.3A–C). For the case in which loss rates decreased as nutrient supply rate increased, the areas with the poorest soils would be dominated by short, leafy, rapidly growing plants with low allocation to roots, whereas areas with the poorest soils would be dominated by less leafy, slowly growing plants with high root allocation on a herbivore-productivity gradient.

There are many other possible trajectories for soil resources and loss rates in natural communities. As already discussed in reviewing Grime (1979), grazed grasslands are likely to have loss rate positively correlated with nutrient supply rate. Each such trajectory would lead to a different pattern of dependence of plant morphologies and life histories on productivity. To make any sense out of such patterns, it is necessary, in studies of patterns in natural vegetation, that both resource supply rates and loss rates be measured, or at least that areas be chosen in which one or the other of these is fairly constant.

QUESTIONS AND CONCLUSIONS

SUMMARY AND SYNTHESIS

This book was motivated by the belief that the similarities that have been observed within and among terrestrial plant communities worldwide could be most parsimoniously explained as resulting from some universal constraints upon plants and from tradeoffs plants face in dealing with these constraints. The major constraint I have focused on is the requirement of plants for both below-ground and above-ground resources—for soil nutrients and light. The physical separation of these essential resources means that plants face an unavoidable tradeoff. To obtain a higher proportion of one resource, a plant must allocate more of its growth to the structures involved in capturing that resource, and thus necessarily would obtain proportionately less of the other resource. This book has explored the logical consequences of these constraints and tradeoffs, and then determined whether the morphological, life history, and community patterns that are predicted to occur because of these constraints and tradeoffs are consistent with those seen in nature. There were striking similarities between the patterns predicted by theory and broad-scale patterns seen in nature.

Plants ranging in life form, stature, and habitat from desert annuals and perennials (Gutierrez and Whitford 1986) to canopy trees in tropical forests (Vitousek 1982), to prairie or savanna or old field grasses and forbs (Tilman 1984, 1987a), to soil algae (personal observation), to douglas fir (Waring et al. 1978), to tundra grasses and shrubs (Shaver and Chapin 1980) have all been found to be resource-limited in their natural habitats. Although different species are superior competitors for different ratios of below-ground resources (Tilman 1982), this book has focused not on this aspect of plant competition, but on the effects of competition for a limiting soil resource versus light. The physical separation of essential plant resources into above-ground and below-ground resources means that a plant structure used to obtain a below-ground resource cannot be used to obtain an above-ground resource, and vice versa. However, because a soil resource and light are nutritionally essential, they are required in a particular proportion for a plant to have balanced, optimal growth. As such, an individual plant faces a tradeoff as it allocates its growth to stems, roots, or leaves. Each allocation pattern, and its resulting morphology, determines the relative amounts of a soil resource and light that a plant can obtain. In combination with a plant's underlying physiology, this constrains each species to being a superior competitor for a narrow range of availabilities of a limiting soil resource and light.

Because each plant species is constrained to being a superior competitor for particular resource levels, the forces that determine resource levels are critically important in determining vegetation patterns. The two major factors determining the average availability of a soil resource and light in a habitat are the supply rate of the soil resource in that habitat and the loss or disturbance rate that the plants experience in that habitat. Each of these leads to a qualitatively different pattern of resource availabilities.

Holding loss rates constant, habitats can be classified by their position along a productivity gradient from habitats with low rates of supply of the limiting soil resource, and thus low plant biomass and high light penetration, to habitats with high rates of supply of the soil resource, and thus high plant biomass and low light penetration. Different plant morphologies, physiologies and life histories are expected at different points along such productivity gradients (Fig. 4.14). Although the exact pattern depends on loss rates, plants dominant on soils with low rates of nutrient supply should have higher allocation to roots and lower allocation to stems than those dominant on soils with more rapid rates of nutrient supply. Patterns consistent with these predictions have been reported for a wide variety of broad-scale climatic, elevational, and soil gradients worldwide (Figs. 5.2–5.5) and also for the much smaller scale productivity gradients that occur within individual forest stands or fields.

If the rate of nutrient supply of a soil were held constant, areas with low loss rates would have the highest plant biomass and those with high loss rates would have the lowest plant biomass. Plant biomass would determine the rate of nutrient uptake and the amount of light intercepted by the plants. Along a gradient from areas with low loss rates to areas with high loss rates, the availability of both soil nutrients and of light at the soil surface would increase, and the height of the canopy would decrease. The higher availabilities of both soil resources and light that occur at higher loss rates, holding soil resource supply rates constant, favor plants that allocate more to leaves and less to roots and shoots. Such plants have higher maximal rates of vegetative growth (*RGR*) because of their higher allocation to leaves. Thus, the availabilities of a limiting soil resource and of light should be positively correlated along a loss rate gradient and be negatively correlated along a productivity gradient. Any actual habitat must be characterized by both its loss rate and by its rate of supply of soil resources, for both of these are critical determinants of the long-term average levels to which plants will reduce a limiting soil resource and light in that habitat.

Many other traits of plants are predicted to depend on the rate of nutrient supply and the loss rate of the habitat. Plant height at maturity, and thus allocation to stem, should increase along productivity gradients and decrease along loss rate gradients. These should be maximal in undisturbed, nutrient-rich habitats, and minimal in highly disturbed, nutrient-poor habitats. Seed size should increase along productivity gradients and decrease along loss rate gradients. Moreover, the same forces that lead to separation of biomes on a broad geographic scale should lead to a similar separation of species along small-scale soil gradients, should select for similar genetic differentiation among individuals within a single population, and should favor similar patterns of morphological and physiological plasticity within each individual. Thus, the pattern of plasticity within individual plants, the pattern of heritable variation among individuals within a given species, the differences among species within a habitat, and the broad-scale separation of species from one habitat to the next are all influenced by the same constraints, the same morphological and physiological tradeoffs, and the same patterns of resource availability.

In addition to influencing the long-term average abundances of plants in habitats, the traits that allow particular plants to persist in a habitat can have a great impact on community dynamics. Plants that are superior competitors in habitats with high loss rates should have higher maximal rates of vegetative growth because of their greater allocation to leaves. These greater maximal growth rates mean that these species will often be the first to attain dominance in a habitat after a disturbance or experimental manipulation. However, they should be competitively displaced by a series of species that are increasingly better competitors for the long-term average resource availabilities determined by the nutrient supply rate and the loss rate of that habitat. Especially on nutrient-rich soils, it may be that most of the dynamics of succession can be explained as a transient effect controlled by the inverse relation between stem allocation and maximal growth rates (see Fig. 4.10). Succession on extremely poor soils, as occurs during primary succession, is expected to share some qualitative features with that on rich soils, but to be determined more by the slow changes in soil nutrient supply rates than by the transient dominance of leafy species. As such, the early dominants of successions on rich soils should be species with low allocations to roots, high allocation to leaves, and low to moderate allocation to stems. In contrast, during primary succession or during secondary succession on a very poor soil, the initial dominants should have high allocation to roots (or to root functions such as nitrogen fixation), and low allocation to stems, with moderate allocation to leaves.

Just as there are morphological constraints and tradeoffs, there are also physiological constraints and tradeoffs. Chapin (1980), for instance, reported that maximal photosynthetic rates increase with leaf protein content. However, protein allocated to photosynthetic machinery cannot be used to increase root absorption capacity or the concentration of secondary compounds that function to reduce herbivory. As discussed in Chapter 4, if all species are assumed to have the same underlying physiology, each unique habitat (i.e., each unique combination of loss or disturbance rates and of nutrient supply rates) should favor a particular plant morphology. If it were assumed that there is a range of plant physiologies determined by patterns of allocation to various biochemical functions, then there should be a correspondence between physiology and morphology. Nutrient-poor, undisturbed habitats would favor plants with efficient root absorption and with high root biomass. Nutrient-rich, undisturbed habitats would favor plants with less efficient roots and/or lower root biomass, but with more efficient photosynthesis and with higher stem and leaf biomass. There is much that needs to be done to elucidate the constraints on and tradeoffs among major physiological processes and between morphological and physiological processes. As these factors become clear, they should be explicitly integrated with the more obvious constraints on and tradeoffs among morphological structures that have been discussed in this book.

Although morphological and physiological plasticity will influence species distributions on productivity and loss rate gradients, the effects of such plasticity would be limited by the presence of physiologically different competitors, structural limitations imposed by the characteristics of the materials from which stems are composed, and physiological limitation imposed by the limited biochemical repertoire of each individual. Optimal foraging for soil resources and light may favor morphological and physiological plasticity within individual plants, but it is unlikely to have a major qualitative effect on the pattern of separation of species along productivity or loss rate gradients. Thus, the universal requirement of terrestrial vascular plants for both above-ground and below-ground resources constrains each species to being a superior competitor for a narrow range of availabilities of these resources. There should be a strong correspondence between the physiology, morphology, and life histories of plants and the habitats in which they are dominant. Although the size structure of plants and the corresponding vertical light gradient means that they can only be an approximation, resource-dependent growth isoclines seem to be a useful summary of the effects of all these aspects of plant biology on plants' ability to compete in a particular habitat.

It was the purpose of this book to explore the logical implications of some of the major mechanisms of plant competition and some fundamental constraints on plant morphology to see how well they might explain patterns we see in nature. There are broad correspondences between the predictions of this theory and natural patterns. Clearly, the theory could be made more realistic and thus more complex. Modifications to this theory that included allocation to compounds or structures that reduced herbivory, or that explicitly included neighborhood aspects of competition, or that allowed plants to be morphologically and physiologically plastic, could all provide significant insights. There is much to be done.

Although the natural world is complex we, as ecologists, cannot afford to revel in that complexity, but must seek explanations—the simplest viable explanations—for the patterns we observe. This book suggests that a few major mechanisms of plant interaction can explain much of the pattern we see in the dynamics and structure of plant communities.

Only careful, long-term experimental and observational studies can determine how useful the ideas presented here actually are. However, from the observational and experimental research reviewed here and in Tilman (1982), it is clear that consumer-resource interactions are a critical factor determining the distribution and abundance of species in both natural and managed populations. Further study of the mechanisms of consumer-resource interactions may allow us to understand many of the forces that led to the evolution of the species we see today, that allow these species to persist, and that give rise to patterns in nature.

APPENDIX

Mathematics of the Model ALLOCATE

Many of the assumptions of the model of competition among size-structured plants were given verbally in Chapter 3. Here I present its mathematical form, as embodied in the simulation model ALLOCATE. Instead of presenting the full FORTRAN code, which occupies 5 pages because of input, output, and housekeeping details, I present the major mathematical elements of the model and discuss their meaning. I do this using the notation common in FORTRAN models, with an * meaning multiplication, and with single variables often being several letters long, often with a mnemonic meaning. For instance, BR means "*b*iomass of *r*oots." I could also express these equations as continuous differential equations, but have chosen not to do so here because the numerical details of solving a model can influence its dynamics, especially when larger time steps are used. Thus, I prefer to present the model in the iterated (discrete time) form.

Each individual plant is assumed to grow vegetatively in a continuous manner with its growth rate determined by the nutrient and light environment it experiences. All plants of a given species or genotype that germinate at a given time are treated as an identical cohort of individuals. All plants, through their consumption (which includes consumption of light via shading) influence resource availabilities, and thus growth is resource-dependent. Let's consider species *I*. Species *I* is composed on distinct cohorts, each of which germinated at a particular time. The Jth cohort of species *I* has the following attributes:

BR(I,J) is the root biomass per individual of cohort J of species *I*.

BL(I,J) is the leaf biomass per individual of cohort J of species *I*.

BS(I,J) is the stem biomass per individual of cohort J of species *I*.

AL(I), AR(I), and AS(I) are the proportional allocations to leaves, roots, and stems, respectively, for all cohorts of species *I*.

QNL(I), QNR(I), QNS(I), and QNSD(I) are the nutrient content per unit biomass of leaf, root, stem, and seed for species *I*.

R(I) is the maximal rate of photosynthesis per unit leaf biomass of species *I*.

RR(I), RS(I), and RL(I) are the per unit biomass respiration rates for roots, stems, and leaves, respectively, for species *I*.

KN(I) and KL(I) are the half-saturation constants for nutrient and light limited photosynthesis, respectively, for species *I*.

N(I,J) is the number of individuals of species *I* in cohort J.

D(I,J) is the death rate experienced by individuals of cohort J in species *I*.

NSEED(I) is the number of seeds of species *I*.

BSEED(I) is the biomass per seed (seed size) for species *I*.

HITE(I,J) is the height of all individuals in cohort J of species *I*.

L(X) is the light intensity (proportion of full sunlight) at a distance X above the soil surface.

S is the availability of the limiting soil resource.

TN is the total amount of all forms of the soil nutrient, and is a major determinant of the nutrient supply rate.

THE SOLUTION PROCEDURE

All plants start their life as seeds. Upon germination, the biomass of a seed (BSEED) is allocated to roots, leaves, and stems according to the following equations, where the new cohort is J:

$$BR(I,J) = AR(I)*BSEED(J)$$
$$BL(I,J) = AL(I)*BSEED(J)$$
$$BS(I,J) = AS(I)*BSEED(J)$$

The height of this newly germinated plant, and the height of all plants, is determined by the stem biomass per individual, BS(I,J), by the relation

$$HITE(I,J) = 4*[2500*BS(I,J)]^{0.5}.$$

The light intensity experienced by this plant is determined by the total leaf biomass at heights greater than the height of this individual. This is based on the assumption that all leaves are born in a single monolayer at the top of the stem. Where X is the height of a plant, let total leaf biomass at greater heights be T(X). Then light penetration to plants of height X is assumed to be a negative exponential function of T(X), with

$$L(X) = \exp(-0.0045*T(X)).$$

The availability of the limiting soil resource is S. S is thus the measurable concentration of available forms of the limiting nutrient in the soil. The effect of a given S on the photosynthetic rate of a plant is assumed to be determined by the ratio of root biomass to leaf biomass of that plant. Thus, let C(I,J) be defined as the effective availability of the soil resource, with

$$C(I,J) = S*BR(I,J)/BL(I,J).$$

Let the amount of photosynthate produced during a time interval of length DT by an individual of cohort J in species *I* be DB(I,J). Note that DT is the time step over which the model is solved. Then,

$$DB(I,J) = MIN\ [(DT*R(I)*BL(I,J)*C(I,J)/(C(I,J) + KN(I)))\ \text{or}$$
$$(DT*R(I)*BL(I,J)*L(HITE(I,J))/(KL(I,J) + L(HITE(I,J))))] - RESP.$$

The presence of the minimum function (MIN) means that the rate of production of new photosynthate per individual is determined by either the nutrient or by light, whichever leads to the lower rate. The biomass produced by each individual of cohort J of species *I* is first used to meet respiratory demands of the plants. This is done by subtracting the respiratory demand from the current production. The respiratory demand, RESP, is determined as

$$RESP = DT*[RR(J)*BR(I,J) + RS(J)*BS(I,J) + RL(J)*BL(I,J)]$$

If $DB(I,J) > 0$, the individual has net production within the time interval, and this production is allocated to roots, leaves, and stems if the individual has not yet reached its maximum height (called sub-adults) or to seeds if it has reached its maximum height (called adults). If $DB(I,J) < 0$, the plant does not grow, but it is still subject to mortality. For sub-adults, the proportion of the new production allocated to leaves, roots, and stems is determined by AL(J), AR(J), and AS(J), respectively. For example, the leaf biomass of an individual would increase by DB(I,J)*AL(J) at the end of this time interval. The new value for the stem biomass per individual of this cohort is then used to calculate its new height. For adults, all excess production is allocated to seeds. The number of seeds produced is simply equal to DB(I,J)/BSEED(J). All of these seeds germinate at the start of the next time interval.

Except in special formulations of the model, seeds experience no mortality. The number of individual plants within a cohort that survive to the end of a time interval (and thus to the start of the next time interval) is determined by

$$SURVIVORS(I,J) = N(I,J)*(1 - DT*D(I,J)),$$

which becomes the new number of individuals in this cohort at the start of the next time interval.

This process of nutrient-dependent growth, allocation, seed production, and survivorship is repeated for all individuals of all cohorts of all species for a given time interval. All these individuals are assumed to experience the same soil nutrient availability and the same vertical light gradient during this time interval. Once all cohorts of all species have been allowed to grow, reproduce, and die during a given time interval, the total effect of their growth on resource availabilities is determined. Light availability is determined as above, using the new heights and new leaf biomass per individual for all the individuals. Nutrient availability, S, is determined as the result of two processes: nutrient uptake and nutrient supply. Nutrient uptake is assumed to be sufficiently great that it just balances the nutrients needed to produce the new biomass or seed of all individual plants. Thus, nutrient uptake by each individual of species *I* and cohort J depends on whether it is an adult or a sub-adult. If it is a sub-adult (producing more leaves, roots, and stems), the uptake per individual would be

$$UPTAKE(I,J) = DB(I,J)*[AR(J)*QNR(J) + AL(J)*QNL(J) + AS(J)*QNS(J)].$$

If it were an adult, its uptake would be determined by the nutrient content of its seeds, as

$$UPTAKE(I,J) = DB(I,J)*QNSD(J).$$

Each of these uptake rates per individual of species *I* and cohort J are multiplied by the number of individuals surviving to the end of the time interval, and these are then summed over all cohorts of all species. This gives the total uptake rate of the entire community. The supply rate of nutrient is calculated as

$$SUPPLY = DT*A*(TN - BN - S),$$

where BN is the total amount of the nutrient that is contained in plant biomass. Thus, nutrient supply is proportional to the proportion of total nutrient that remains in the soil.

Given the total uptake and the total supply amounts for the time interval of length DT, the new availability of the nutrient is just equal to the former availability (S) plus the SUPPLY minus the total uptake.

After this process has been completed one time, it is repeated with the new values for all variables used as the new starting conditions. In summary, then, this is a model of growth and competition among size-structured plants. Plant growth rates are determined by nutrient and light availabilities. By their growth, the plants produce a vertical light gradient. The process of growth and nutrient consumption is solved iteratively, with plants first growing in response to current resource levels, and thus creating the plant sizes and number of individuals for all cohorts of all species for the next iteration. After this is done, the effects of this growth on nutrient availability and on the vertical light gradient are determined. These new values then determine the pattern of growth for the next iteration.

Because seeds were allowed to germinate immediately after they were produced, a new cohort of plants could be produced every day. This caused there to be too large a number of cohorts to allow the model to be solved numerically, even with a Cray 2. To overcome this problem, I combined similar cohorts by establishing a limit of 40 cohorts per species. Every 20 iterations, if a species had more than 20 different cohorts, I determined whether any of these cohorts differed by less than 1/40 of the possible range in biomass per individual for that species. Those that did differ by less than 1/40 were combined.

I solved the model numerically using CRAY FORTRAN on a Cray 2 supercomputer. For most simulations, I used a time step of about one day (actually 0.7 day). With a maximal growth rate of 0.2 day^{-1} for a plant that allocated 95% of its mass to leaves, this gave sufficiently small iterations that solutions were basically time-step-size independent. For all simulations, a period of 140 days was considered to constitute a single growing season or "year." Although ALLOCATE must be solved on a supercomputer when hundreds of species compete, versions in which from 2 to about 20 species compete can be solved on microcomputers, such as the IBM-XT. Copies of the full FORTRAN code of ALLOCATE, as formulated for a microcomputer, are available upon request.

References

Aarsen, L. W., and R. Turkington. 1985. Within-species diversity in natural populations of *Holcus lanatus*, *Lolium perenne* and *Trifolium repens* from four different-aged pastures. *Journal of Ecology* 73:869-886.

Albrektson, A., A. Aronsson, and C. O. Tamm. 1977. The effect of forest fertilization on primary production and nutrient cycling in the forest ecosystem. *Silva Fenn.* 11:233-239.

Armstrong, R. A., and R. McGehee. 1980. Competitive exclusion. *American Naturalist* 115:151-170.

Baker, H. G. 1972. Seed weight in relation to environmental condition in California. *Ecology* 53:997-1010.

Barbour, M., J. Burk, and W. Pitts. 1980. *Terrestrial Plant Ecology*. Menlo Park, CA: Benjamin/Cummings Publishing Company.

Bazzaz, F. A. 1979. The physiological ecology of plant succession. *Annual Review of Ecology and Systematics* 10:351-371.

Beals, E. W., and J. B. Cope. 1964. Vegetation and soils in an eastern Indiana woods. *Ecology* 45:777-792.

Beard, J. S. 1944. Climax vegetation in tropical America. *Ecology* 25:127-158.

Beard, J. S. 1955. The classification of tropical American vegetation types. *Ecology* 36:89-100.

Beard, J. S. 1983. Ecological control of the vegetation of Southwestern Australia: moisture versus nutrients. In F. J. Kruger, D. T. Mitchell and J.V.M. Jarvis, eds., *Mediterranean-Type Ecosystems*, 66-73. New York: Springer-Verlag.

Bender, E. A., T. J. Case, and M. E. Gilpin. 1984. Perturbation experiments in community ecology: theory and practice. *Ecology* 65:1-13.

Billings, W. D. 1938. The structure and development of old field shortleaf pine stands and certain associated physical properties of the soil. *Ecological Monographs* 8:437-499.

Black, J. N. 1958. Competition between plants of different initial seed sizes in swards of subterranean clover (*Trifolium subterraneum* L.) with particular reference to leaf area and the light microclimate. *Australian Journal of Agricultural Research* 9:299-318.

Black, J. N. 1960. An assessment of the role of planting density in comparison between red clover (*Trifolium pratense* L.) and lucerne (*Medicago sativa* L.) in the early vegetative stage. *Oikos* 11:26-42.

Bloom, A. J., F. S. Chapin III, and H. A. Mooney. 1985. Resource limitation in plants—an economic analogy. *Annual Review of Ecology and Systematics* 16:363-392.

Bormann, F. H., and G. E. Likens. 1979. *Pattern and Process in a Forested Ecosystem.* New York: Springer-Verlag.

Box, T. W. 1961. Relationships between plants and soils of four range plant communities in south Texas. *Ecology* 42:794-810.

Brenchley, W., and K. Warington. 1958. *The Park Grass Plots at Rothamsted.* Harpendon, U.K.: Rothamsted Experimental Station.

Brix, H. 1983. Effects of thinning fertilization on growth of Douglas fir: relative contribution of foliage quantity and efficiency. *Canadian Journal of Forest Research* 13:167-175.

Brown, J. H., D. W. Davidson, J. C. Munger, and R. S. Inouye. 1986. Experimental community ecology: the desert granivore system. In J. Diamond and T. J. Case, eds., *Community Ecology*, 41-62. New York: Harper and Row.

Chabot, B. F., and H. A. Mooney (eds.). 1985. *Physiological Ecology of North American Plant Communities.* London: Chapman and Hall.

Chapin, F. S. III. 1980. The mineral nutrition of wild plants. *Annual Review of Ecology and Systematics* 11:233-260.

Chapin, F. S. III, and G. R. Shaver. 1985. Arctic. In B. F. Chabot and H. A. Mooney, eds., *Physiological Ecology of North American Plant Communities*, 16-40. London: Chapman and Hall.

Charnov, E. L., and W. M. Schaffer. 1973. Life-history consequences of natural selection: Cole's result revisited. *American Naturalist* 107:791-793.

Chichester, F. W., J. O. Legg, and G. Stanford. 1975. Relative mineralization rates of indigenous and recently incorporated N-labelled nitrogen. *Soil Science* 120:455-460.

Christensen, N. L., and R. K. Peet. 1981. Secondary forest succession on the North Carolina Piedmont. In D. C. West, H. H. Shugart, and D. B. Botkin, eds., *Forest Succession: Concepts and Applications*, 230-244. New York: Springer-Verlag.

Clapham, A. R., T. G. Tutin, and E. F. Warburg. 1962. *Flora of the British Isles*, 2nd ed. Cambridge: Cambridge University Press.

Clatworthy, J. N. 1960. Studies on the nature of competition between closely related species. Ph.D. Thesis, University of Oxford.

Clements, F. E. 1916. *Plant Succession*. Carnegie Institute Washington Publication 242. 512 pages.

Cody, M. L. 1986. Structural niches in plant communities. In J. Diamond and T. Case, eds., *Community Ecology*, 381-405. New York: Harper and Row.

Cody, M. L., and H. A. Mooney. 1978. Convergence versus nonconvergence in Mediterranean-climate ecosystems. *Annual Review of Ecology and Systematics* 9:265-321.

Cole, L. C. 1954. The population consequences of life history phenomena. *Quarterly Review of Biology* 29:103-137.

Connell, J. 1983. On the prevalence and relative importance of interspecific competition: evidence from field experiments. *American Naturalist* 122:661-696.

Connolly, J. 1986. On difficulties with replacement-series

methodology in mixture experiments. *Journal of Applied Ecology* 23:125-137.

Cooper, A. W. 1981. Above-ground biomass accumulation and net primary production during the first 70 years of succession in *Populus grandidentata* stands on poor sites in northern lower Michigan. In D. C. West, H. H. Shugart, and D. B. Botkin, eds., *Forest Succession: Concepts and Applications*, 339-360. New York: Springer-Verlag.

Cooper, W. S. 1913. The climax forest of Isle Royale, Lake Superior, and its development. *Botanical Gazette* 55:1-44, 115-140, 189-235.

Cooper, W. S. 1923. The recent ecological history of Glacier Bay, Alaska. *Ecology* 4:93-128, 223-246, 355-365.

Cooper, W. S. 1939. A fourth expedition to Glacier Bay, Alaska. *Ecology* 20:130-155.

Covich, A. 1972. Ecological economics of seed consumption by *Peromyscus*—a graphical model of resource substitution. *Transactions of the Connecticut Academy of Arts and Science* 44:71-93.

Cowan, M. L. 1986. Growth responses of old-field plants to a nitrogen gradient. M.S. Thesis, University of Minnesota.

Cowles, H. C. 1899. The ecological relations of the vegetation on the sand dunes of Lake Michigan. *Botanical Gazette* 27:95-117, 167-202, 281-308, 361-391.

Crocker, R. L., and B. A. Dickson. 1957. Soil development on the recessional moraines of the Herbert and Mendenhall glaciers of southeastern Alaska. *Journal of Ecology* 45:169-185.

Crocker, R. L., and J. Major. 1955. Soil development in relation to vegetation and surface age at Glacier Bay, Alaska. *Journal of Ecology* 43:427-448.

de Wit, C. T. 1960. On competition. *Agricultural Research Reports (Versl. landbouwk. Onderz.)* 66.8, Wageningen, The Netherlands. 88 pages.

Donald, C. M. 1951. Competition among pasture plants. I.

Intra-specific competition among annual pasture plants. *Australian Journal of Agricultural Research* 2:355-376.

Farrow, E. P. 1916. On the ecology of the vegetation of Breckland. II. *Journal of Ecology* 4:57-64.

Farrow, E. P. 1917. On the ecology of the vegetation of Breckland. IV. *Journal of Ecology* 5:104-113.

Fitter, A. H. 1986. Acquisition and utilization of resources. In M. J. Crawley, ed., *Plant Ecology*, 375-405. Oxford: Blackwell Scientific Publications.

Fowells, H. A. 1965. *Silvics of Forest Trees of the United States*. U.S. Department of Agriculture, Washington, D.C.

Fowler, N. 1981. Competition and coexistence in a North Carolina grassland. II. The effects of the experimental removal of species. *Journal of Ecology* 69:843-854.

Fox, J. F. 1977. Alternation and coexistence of tree species. *American Naturalist* 111:68-89.

Gifford, R. M., and C. L. Jenkins. 1981. Prospects of applying knowledge of photosynthesis toward improving crop production. In Gorindjee, ed., *Photosynthesis: CO_2 Assimilation and Plant Productivity*, Vol. 2. New York: Academic Press.

Givnish, T. J. 1982. On the adaptive significance of leaf height in forest herbs. *American Naturalist* 120:353-381.

Gleason, H. A. 1917. The structure and development of plant association. *Bulletin of the Torrey Botany Club* 44:463-481.

Gleason, H. A. 1927. Further views on the succession concept. *Ecology* 8:299-326.

Gleason, H. A., and A. Cronquist. 1963. *Manual of Vascular Plants of Northeastern United States and Adjacent Canada*. New York: D. Van Nostrand Company.

Godwin, H. 1956. *The History of the British Flora*. Cambridge: Cambridge University Press.

Greenhill, G. 1881. Determination of the greatest height consistent with stability that a vertical pole or mast can be made, and of the greatest height to which a tree of given proportions can grow. *Proceedings of the Cambridge Philosophical Society* 4:65-73.

Grigal, D. F., L. M. Chamberlain, H. R. Finney, D. V. Wroblewski, and E. R. Gross. 1974. *Soils of the Cedar Creek Natural History Area.* Miscellaneous Report No. 123. St. Paul, MN: University of Minnesota Agricultural Experimental Station.

Grime, J. P. 1979. *Plant Strategies and Vegetation Processes.* Chichester: John Wiley & Sons.

Grime, J. P., and R. Hunt. 1975. Relative growth rate: its range and adaptive significance in a local flora. *Journal of Ecology* 63:393-422.

Grime, J. P., and D. W. Jeffery. 1964. Seedling establishment in vertical gradients of sunlight. *Journal of Ecology* 53:621-642.

Grimm, E. C. 1984. Fire and other factors controlling the big woods vegetation of Minnesota in the mid-nineteenth century. *Ecological Monographs* 54:291-311.

Gutierrez, J. R., and W. G. Whitford. 1986. Responses of Chihuahuan Desert herbaceous annuals to rainfall augmentation. *Journal of Arid Environments,* in press.

Hairston, N. G., F. E. Smith, and L. B. Slobodkin. 1960. Community structure, population control, and competition. *American Naturalist* 94:421-425.

Hanawalt, R. B., and R. H. Whittaker. 1976. Altitudinally coordinated patterns of soils and vegetation in the San Jacinto Mountains, California. *Soil Science* 121:114-124.

Harper, J. L. 1961. Approaches to the study of plant competition. In F. L. Milthorpe, ed., *Mechanisms in Biological Competition. Symp. Soc. exp. Biol.* 15:1-39.

Harper, J. L. 1977. *Population Biology of Plants.* London: Academic Press.

Harper, J. L., and J. Ogden. 1970. The reproductive strategy of higher plants. 1. The concept of strategy with special reference to *Senecio vulgaris* L. *Journal of Ecology* 58:681-698.

Harris, G. P. 1986. *Phytoplankton Ecology: Structure, Function and Fluctuation*. London: Chapman and Hall.

Hole, F. D. 1976. *Soils of Wisconsin*. Madison: University of Wisconsin Press.

Horn, H. S. 1971. *The Adaptive Geometry of Trees*. Princeton: Princeton University Press.

Howrath, S. E., and J. T. Williams. 1972. *Chrysanthemum segetum*. In "Biological Flora of the British Isles," *Journal of Ecology* 60:573-584.

Hsu, S. B., S. P. Hubbell, and P. Waltman. 1977. A mathematical theory for single-nutrient competition in continuous cultures of microorganisms. *SIAM Journal of Applied Mathematics* 32:366-383.

Hubbell, S. P., and R. B. Foster. 1986. Biology, chance, and history and the structure of tropical rain forest tree communities. In J. Diamond and T. J. Case, eds., *Community Ecology*, 314-330. New York: Harper and Row.

Hubbell, S. P., and P. A. Werner. 1979. On measuring the intrinsic rate of increase of populations with heterogeneous life histories. *American Naturalist* 113:277-293.

Inouye, R. S. 1980. Density-dependent germination response by seeds of desert annuals. *Oecologia* 46:235-238.

Inouye, R. S., G. S. Byers, and J. H. Brown. 1980. Effects of predation and competition on survivorship, fecundity, and community structure of desert annuals. *Ecology* 61:1344-1351.

Inouye, R. S., and W. M. Schaffer. 1981. On the ecological meaning of ratio (de Wit) diagrams in plant ecology. *Ecology* 62:1679-1681.

Inouye, R. S., N. J. Huntly, D. Tilman, J. R. Tester, M. A. Stillwell, and K. C. Zinnel. 1987a. Old field succession on a Minnesota sand plain. *Ecology* 68:12-26.

Inouye, R. S., N. J. Huntly, D. Tilman, and J. R. Tester. 1987b. Pocket gophers, vegetation and soil nitrogen along a succession sere in east central Minnesota. *Oecologia* 72:178-184.

Iwasa, Y., and J. Roughgarden. 1984. Shoot/root balance of plants: optimal growth of a system with many vegetative organs. *Theoretical Population Biology* 25:78-104.

Janzen, D. H. 1970. Herbivores and the number of trees in tropical forests. *American Naturalist* 104:501-528.

Jarvis, P. G., and M. S. Jarvis. 1964. Growth rates of woody plants. *Physiologia Pl.* 17:654-666.

Jenny, H. 1980. Soil genesis with ecological perspectives. *Ecological Studies*, vol. 37. New York: Springer-Verlag.

Kira, T. 1964. Metabolic aspects of the tropical rain forest. *Nature (Shizen)* 19:22-29.

Kira, T., H. Ogawa, and K. Shinozaki. 1953. Intraspecific competition among higher plants. 1. Competition-density-yield inter-relationships in regularly dispersed populations. *J. Inst. Polytech. Osaka Cy. Univ. D.* 4:1-16.

Kuriowa, S. 1960. Intraspecific competition in artificial sunflower community. *Bot. Mag. Tokyo* 73: 300-309.

Lassoie, J. P., T. M. Hinckley, and C. C. Grier. 1985. Coniferous forests in the Pacific Northwest. In B. F. Chabot and H. A. Mooney, eds., *Physiological Ecology of North American Plant Communities*, 127-161. London: Chapman and Hall.

Lawes, J., and J. Gilbert. 1880. Agricultural, botanical and chemical results of experiments on the mixed herbage of permanent grassland, conducted for many years in succession on the same land. I. *Philosophical Transactions of the Royal Society* 171:189-514.

Lawlor, L. R. 1979. Direct and indirect effects of n-species competition. *Oecologia* 43:355-364.

Lawrence, D. B. 1958. Glaciers and vegetation in Southeastern Alaska. *American Scientist* 46:89-122.

Lawrence, D. B. 1979. Primary versus secondary succession at Glacier Bay National Monument, southeastern Alaska. In R. M. Linn, ed., *Proceedings of the First Conference on Scientific Research in the National Parks*, 213-224.

Lawrence, D. B., R. E. Schoenike, A. Quispel, and G. Bond. 1967. The role of *Dryas drummondi* in vegetation development following ice recession at Glacier Bay, Alaska, with special reference to its nitrogen fixation by root nodules. *Journal of Ecology* 55:793-813.

Levine, S. H. 1976. Competitive interactions in ecosystems. *American Naturalist* 110:903-910.

Likens, G. E., F. H. Bormann, R. S. Pierce, J. S. Eaton, and N. M. Johnson. 1977. *Bio-geo-chemistry of a Forested Ecosystem*. New York: Springer-Verlag.

Lindsey, A. A. 1961. Vegetation of the drainage-aeration classes of northern Indiana soils in 1830. *Ecology* 42:432-436.

Loach, K. 1970. Shade tolerance in tree seedlings: II. Growth analysis of plants raised under artificial shade. *New Phytologist* 69:273-286.

Lubchenco, J. 1978. Plant species diversity in a marine intertidal community: importance of herbivore food preference and algal competitive abilities. *American Naturalist* 112:23-39.

MacMahon, J. A. 1981. Successional processes: comparisons among biomes with special reference to probable roles of and influences on animals. In D. C. West, H. H. Shugart, and D. B. Botkin, eds., *Forest Succession: Concepts and Applications*, 227-304. New York: Springer-Verlag.

Mahmoud, A., and J. P. Grime. 1976. An analysis of competitive ability in three perennial grasses. *New Phytologist* 77:431-435.

Marks, P. L. 1983. On the origin of the field plants of the northeastern United States. *American Naturalist* 122:210-228.

May, R. M. 1973. *Stability and Complexity in Model Ecosystems*. Princeton: Princeton University Press.

McMahon, T. A. 1973. Size and shape in biology. *Science* 179:1201-1204.

Melillo, J. M., J. D. Aber, and J. F. Muratore. 1982. Nitrogen and lignin control of hardwood leaf litter decomposition dynamics. *Ecology* 63:621-626.

Miller, H. G., and J. D. Miller. 1976. Effect of nitrogen supply on net primary production in Corsican pine. *Journal of Applied Ecology* 13:249-256.

Milton, W. 1947. The yield, botanical and chemical composition of natural hill herbage under manuring, controlled grazing and hay conditions. I. Yield and botanical. *Journal of Ecology* 35:65-89.

Monsi, M. 1968. Mathematical models of plant communities. In F. E. Eckardt, ed., *Functioning of Terrestrial Ecosystems at the Primary Production Level*, 131-149. Liege, Belgium: Vaillant-Carmanne, S.A.

Mooney, H. A. 1972. The carbon balance of plants. *Annual Review of Ecology and Systematics* 3:315-346.

Mooney, H. A. 1977. *Convergent Evolution in Chile and California*. Stroudsburg, PA: Dowden, Hutchinson & Ross.

Newberry, D. M., and E. I. Newman. 1978. Competition between grassland plants of different initial sizes. *Oecologia* 33:361-380.

O'Brien, W. J. 1974. The dynamics of nutrient limitation of phytoplankton algae: a model reconsidered. *Ecology* 50:930-938.

Odum, E. P. 1960. Organic production and turnover in old field succession. *Ecology* 41:34-49.

Oksanen, L., S. D. Fretwell, J. Arrud, and P. Niemala. 1981. Exploitation ecosystems in gradients of primary productivity. *American Naturalist* 118:240-261.

Olson, J. S. 1958. Rates of succession and soil changes on southern Lake Michigan sand dunes. *Botanical Gazette* 119:125-169.

Orians, G. H., and R. T. Paine. 1983. Convergent evolution at the community level. In D. J. Futuyma and M. Slatkin, eds., *Coevolution*, 431-458. Sutherland, MA: Sinauer Associates.

Pacala, S. W. 1986. Neighborhood models of plant population dynamics. 4. Single-species and multispecies models of annuals with dormant seeds. *American Naturalist* 128:859-878.

Pacala, S. W., and J. A. Silander, Jr. 1985. Neighborhood models of plant population dynamics. 1. Single-species models of annuals. *American Naturalist* 125:385-411.

Parrish, J.A.D., and F. A. Bazzaz. 1982. Responses of plants from three successional communities to a nutrient gradient. *Journal of Ecology* 70:233-248.

Pastor, J., J. D. Aber, C. A. McClaugherty, and J. M. Melillo. 1982. Geology, soils, and vegetation of Blackhawk Island, Wisconsin. *American Midland Naturalist* 108:266-277.

Pastor, J., J. D. Aber, C. A. McClaugherty, and J. M. Melillo. 1984. Aboveground production and N and P cycling along a nitrogen mineralization gradient on Blackhawk Island, Wisconsin. *Ecology* 65:256-268.

Pastor, J., M. A. Stillwell, and D. Tilman. 1987. Nitrogen mineralization and nitrification in four Minnesota old fields. *Oecologia* 71:481-485.

Pigott, C. D., and K. Taylor. 1964. The distribution of some woodland herbs in relation to the supply of nitrogen and phosphorus in the soil. *Journal of Ecology* 52 (supplement):175-185.

Quinn, J. F., and A. E. Dunham. 1983. On hypothesis testing in ecology and evolution. *American Naturalist* 122:602-617.

Rabinovitch-Vin, A. 1979. Influence of parent rock on soil properties and composition of vegetation in the Galilee. Ph.D. Thesis. Hebrew University of Jerusalem.

Rabinovitch-Vin, A. 1983. Influence of nutrients on the composition and distribution of plant communities in Mediterranean-Type ecosystems of Israel. In F. J. Kruger, D. T. Mitchell, and J.V.M. Jarvis, eds., *Mediterranean-Type Ecosystems*, 74-85. New York: Springer-Verlag.

Rapport, D. J. 1971. An optimization model of food selection. *American Naturalist* 105:575-578.

Rapport, D. J., and J. E. Turner. 1975. Feeding rates and population growth. *Ecology* 56:942-949.

Raunkiaer, C. 1934. *The Life Forms of Plants and Statistical Plant Geography*. London: Oxford Press.

Reiners, W. A., I. A. Worley, and D. B. Lawrence. 1971. Plant diversity in a chronosequence at Glacier Bay, Alaska. *Ecology* 52:55-69.

Rice, E. L., W. T. Penfound, and L. M. Rohrbaugh. 1960. Seed dispersal and mineral nutrition in succession in abandoned fields in central Oklahoma. *Ecology* 41:224-228.

Ricklefs, R. E. 1987. Community diversity: relative roles of local and regional process. *Science* 235:167-171.

Robertson, G. P. 1982. Factors regulating nitrification in primary and secondary succession. *Ecology* 63:1561-1573.

Robertson, G. P., and P. M. Vitousek. 1981. Nitrification potentials in primary and secondary succession. *Ecology* 62:376-386.

Salisbury, E. J. 1942. *The Reproductive Capacity of Plants*. London: G. Bell and Sons, Ltd.

Schaffer, W. M. 1981. Ecological abstraction: the consequences of reduced dimensionality in ecological models. *Ecological Monographs* 51:383-401.

Schoener, T. W. 1983. Field experiments on interspecific competition. *American Naturalist* 122:240-285.

Schopf, J. W., J. M. Hayes, and M. R. Walter. 1983. Evolution of earth's earliest ecosystems: recent progress and unsolved problems. In J. William Schopf, ed., *Earth's Earliest Biosphere: Its Origin and Evolution*, 361-384. Princeton: Princeton University Press.

Shamsi, S.R.A., and F. H. Whitehead. 1977. Comparative eco-physiology of *Epilobium hirsutum* L. and *Lythrum salicaria* L. III. Mineral nutrition. *Journal of Ecology* 65:55-70.

Shaver, G. R., and F. S. Chapin, III. 1980. Response to fertilization by various plant growth forms in an Alaskan tundra: nutrient accumulation and growth. *Ecology* 61:662-675.

Snaydon, R. W. 1962. Micro-distribution of *Trifolium repens* L. and its relation to soils factors. *Journal of Ecology* 50:133-143.

Snaydon, R. W. 1970. Rapid population differentiation in a mosaic environment. I. The response of *Anthoxanthum odoratum* populations to soils. *Evolution* 24:257-269.

Snaydon, R. W. 1976. Genetic change within species. In J. M. Thurston, G. V. Dyke and E. D. Williams, *The Park Grass Experiment*, 14-20. Harpendon, U.K.: Rothamsted Experimental Station.

Snaydon, R. W., and M. S. Davies. 1972. Rapid population differentiation in a mosaic environment. II. Morphological variation in *Anthoxanthum odoratum*. *Evolution* 26:390-405.

Sonleitner, F. J. 1977. A stochastic computer model for simulating population growth. *Researches on Population Ecology* 19:10-32.

Spitters, C.J.T. 1983. An alternative approach to the analysis of mixed cropping experiments. 1. Estimation of competition effects. *Netherlands Journal of Agricultural Science* 31:1-11.

Stebbins, G. L., and G.J.C. Hill. 1980. Did multicellular plants invade the land? *American Naturalist* 115:342-353.

Stern, W. R., and C. M. Donald. 1962. Light relationships in grass-clover swards. *Australian Journal of Agricultural Research* 13:599-614.

Summerhayes, V. 1941. Effects of voles (*Microtus agrestis*) on vegetation. *Journal of Ecology* 29:14-48.

Swank, W. T., J. B. Waide, D.A. Crossley, Jr., and R. L. Todd. 1981. Insect defoliation enhances nitrate export from forest ecosystems. *Oecologia* (Berlin) 51:297-299.

Tamm, C. O. 1985. The Swedish optimum nutrition experiments in forest stands—aims, methods and yield results.

Journal of the Royal Swedish Academy of Agriculture and Forestry, Suppl. 17: 9-29.

Tamm, C. O., and A. Aronsson. 1982. Optimum nutrition of some non-food plants. Pages 181-196 in *Optimizing Yields—The Role of Fertilizers*, Proceedings of the 12th Congr. Int. Potash Institute, Bern.

Tamm, C. O., A. Aronsson, and H. Burgtorf. 1974. The optimum nutrition experiment Strasan. A brief description of an experiment in a young stand of Norway spruce (*Picea abies* Karst.). *Rapporter och Uppsatser*, No. 17, Dept. of Plant Ecology and Forest Soils. Royal College of Forestry, Stockholm, Sweden. 29 pp.

Tansley, A. G. 1949. *The British Isles and Their Vegetation*. Cambridge: Cambridge Press.

Thurston, J. 1969. The effect of liming and fertilizers on the botanical composition of permanent grassland, and on the yield of hay. In I. Rorison, ed., *Ecological Aspects of the Mineral Nutrition of Plants*, 3-10. Oxford: Blackwell Scientific Publications.

Tilman (Titman), D. 1976. Ecological competition between algae: experimental confirmation of resource-based competition theory. *Science* 192:463-465.

Tilman, D. 1980. Resources: a graphical-mechanistic approach to competition and predation. *American Naturalist* 116:362-393.

Tilman, D. 1982. *Resource Competition and Community Structure*. Princeton: Princeton University Press.

Tilman, D. 1983. Plant succession and gopher disturbance along an experimental gradient. *Oecologia* 60:285-292.

Tilman, D. 1984. Plant dominance along an experimental nutrient gradient. *Ecology* 65:1445-1453.

Tilman, D. 1985. The resource ratio hypothesis of succession. *American Naturalist* 125:827-852.

Tilman, D. 1986a. Evolution and differentiation in terrestrial plant communities: the importance of the soil resource:light gradient. In J. Diamond and T. Case, eds., *Community Ecology*, 359-380. New York: Harper and Row.

Tilman, D. 1986b. Nitrogen-limited growth in plants from different successional stages. *Ecology* 67:555-563.

Tilman, D. 1987a. Secondary succession and the pattern of plant dominance along experimental nitrogen gradients. *Ecological Monographs* 57:189-214.

Tilman, D. 1987b. The importance of the mechanisms of interspecific interaction. *American Naturalist* 129:769-774.

Toumey, J. W. and R. Kienholz, 1931. Trenched plots under forest canopies. Bull. No. 30, School of Forestry, Yale University, New Haven, Connecticut. 31 pp.

Vandermeer, J. H. 1980. Indirect mutualism: Variations on a theme by Stephen Levine. *American Naturalist* 116:441-448.

Vitousek, P. 1982. Nutrient cycling and nutrient use efficiency. *American Naturalist* 119:553-572.

Vitousek, P. M., J. R. Gosz, C. C. Grier, J. M. Melillo, W. A. Reiners, and R. L. Todd. 1979. Nitrate losses from disturbed ecosystems. *Science* 204: 469-474.

Vitousek, P. M., J. R. Gosz, C. C. Grier, J. M. Melillo, and W. A. Reiners. 1982. A comparative analysis of potential nitrification and nitrate mobility in forest ecosystems. *Ecological Monographs* 52:155-177.

Vitousek, P. M., and P. A. Matson. 1985. Disturbance, nitrogen availability, and nitrogen losses in an intensively managed loblolly pine plantation. *Ecology* 66: 1360-1376.

Walker, J., C. H. Thompson, I. F. Fergus, and B. R. Tunstall. 1981. Plant succession and soil development in coastal sand dunes of subtropical eastern Australia. In D. C. West, H. H. Shugart, and D. B. Botkin, eds., *Forest Succession: Concepts and Applications*, 107-131. New York: Springer-Verlag.

Waller, D. M. 1986. The dynamics of growth and form. In M. J. Crawley, ed., *Plant Ecology*, 291-320. Oxford: Blackwell Scientific Publications.

Walter, H. 1985. *Vegetation of the Earth and Ecological Systems of the Geo-biosphere*, 3rd ed. New York: Springer-Verlag.

Waring, R. H., W. H. Emmington, H. L. Gholz, and C. C.

Grier. 1978. Variation in maximum leaf area of coniferous forests in Oregon and its ecological significance. *Forestry Science* 24:131-140.

Werner, P. A., and W. J. Platt. 1976. Ecological relationship of co-occurring goldenrods (Solidago:Compositae). *American Naturalist* 110:959-971.

West, D. C., H. H. Shugart, and D. B. Botkin (eds.). 1981. *Forest Succession: Concepts and Applications*. New York: Springer-Verlag.

Westoby, M. 1984. The self-thinning rule. *Advances in Biological Research* 14:167-225.

White, A. S. 1983. The effects of 13 years of annual prescribed burning on a *Quercus ellipsoidalis* community in Minnesota. *Ecology* 64:1081-1085.

Whitney, G. 1986. Relation of Michigan's presettlement pine forests to substrate and disturbance history. *Ecology* 67:1548-1559.

Whittaker, R. H. 1953. A consideration of climax theory: the climax as a population and pattern. *Ecological Monographs* 23:41-78.

Whittaker, R. H. 1975. *Communities and Ecosystems*. New York: Macmillan Press.

Wiens, J. A. 1984. Resource systems, populations, and communities. In P. W. Price, W. S. Gaud, and C. N. Slobodchikoff, eds., *A New Ecology*, 397-436. New York: Wiley Interscience.

Willis, A. 1963. Braunton Burrows: the effects on the vegetation of the addition of mineral nutrients to the dune soils. *Journal of Ecology* 51:353-374.

Wilson, S. D., and P. A. Keddy. 1986. Species competitive ability and position along a natural stress/disturbance gradient. *Ecology* 67:1236-1242.

Woods, K. D. 1979. Reciprocal replacement and the maintenance of codominance in a beech-maple forest. *Oikos* 33:31-39.

Woods, K. D., and R. H. Whittaker. 1981. Canopy-understory interaction and the internal dynamics of mature hardwood and hemlock-hardwood forests. In D. C. West, H. H. Shugart, and D. B. Botkin, eds., *Forest Succession: Concepts and Applications*, 305- 322. New York: Springer-Verlag.

Worley, I. A. 1973. The "black crust" phenomenon in upper Glacier Bay, Alaska. *Northwest Science* 47:20-29.

Zak, D. R., K. S. Pregitzer, and G. E. Host. 1986. Landscape variation in nitrogen mineralization and nitrification. *Canadian Journal of Forest Research* 16:1258-1263.

Zedler, J., and P. Zedler. 1969. Association of species and their relationship to microtopography within old fields. *Ecology* 50:432-442.

Author Index

Subject Index

MONOGRAPHS IN POPULATION BIOLOGY

EDITED BY ROBERT M. MAY

(continued)

21. Natural Selection in the Wild, by John A. Endler
22. Theoretical Studies on Sex Ratio Evolution, by Samuel Karlin and Sabin Lessard
23. A Hierarchical Concept of Ecosystems, by R. V. O'Neill, D. L. DeAngelis, J. B. Waide, and T.F.H. Allen
24. Population Ecology of the Cooperatively Breeding Acorn Woodpecker, by Walter D. Koenig and Ronald L. Mumme
25. Population Ecology of Individuals, by Adam Łomnicki
26. Plant Strategies and the Dynamics and Structure of Plant Communities, by David Tilman

LIBRARY OF CONGRESS CATALOGING-IN-PUBLICATION DATA

Tilman, David, 1949–
Plant strategies and the dynamics and structure of plant communities.

(Monographs in population biology ; 26)
Bibliography: p.
Includes indexes.
1. Vegetation dynamics. 2. Plant communities.
I. Title. II. Title: Dynamics and structure of plant communities. III. Series.
QK910.T55 1988 581.5′247 87–25833
ISBN 0–691–08488–2 (alk. paper)
ISBN 0–691–08489–0 (pbk. : alk. paper)